GUIDE TO
SCIENTIFIC COMPUTING

SECOND EDITION

GUIDE TO
SCIENTIFIC COMPUTING

SECOND EDITION

PETER R. TURNER

Mathematics Department
United States Naval Academy
Annapolis, MD

MACMILLAN

CRC Press
Boca Raton   London   New York   Washington, D.C.

Published in North America by CRC Press LLC, 2000 Corporate Blvd N.W., Boca Raton, FL 33431, USA

Published in North America under license from Macmillan Press Ltd, Houndmills, Basingstoke, Hants, RG21 6XS, United Kingdom.

Library of Congress Cataloging-in-Publication Data

Catalog record is available from the Library of Congress.

© 2001 by Peter R. Turner

No claim to original U.S. Government works
International Standard Book Number Macmillan Press Ltd. 0-333-79450-8
International Standard Book Number CRC Press LLC 0-8493-1242-6
Printed in the United States of America 1 2 3 4 5 6 7 8 9 0
Printed on acid-free paper

Contents

Preface

This book is a revision of the first edition *Guide to Numerical Analysis,* which was developed from a variety of introductory courses in numerical methods and analysis that I taught at the University of Lancaster and the University of Maryland. Over the intervening ten years or so, I have taught similar courses at the Naval Academy. The level and content of the courses has varied considerably but they have included all the material presented here. The intention of this book is to provide a gentle and sympathetic introduction to many of the problems of scientific computing, and the wide variety of methods used for their solution. The book is therefore suitable for a first course in numerical mathematics or scientific computing – whether it be for mathematics majors, or for students of science, engineering or economics, for example.

This, of course, precludes the possibility of providing a fully rigorous treatment of all the most up-to-date algorithms and their analyses. The intention is rather to give an appreciation of the need for numerical methods for the solution of different types of problem, and to discuss the basic approaches. For each of the problems this is followed by at least some mathematical justification – and, most importantly, examples – to provide both practical evidence and motivation for the reader. The level of mathematical justification is determined largely by the desire to keep the mathematical prerequisites to a minimum. Thus, for example, no knowledge of linear algebra is assumed beyond the most basic matrix algebra, and analytic results are based on a sound knowledge of the calculus.

Inevitably this means that some methods, such as those for numerical solution of ordinary differential equations, are not derived in a detailed and rigorous manner. Such methods are nonetheless justified and/or derived by appealing to other techniques discussed earlier in the book such as those for numerical integration and differentiation. In contrast, in situations where rigorous explanation can be included without substituting bafflement for enlightenment, this is given. In Chapter 2, on iterative solution of equations, for instance, a detailed analysis of the convergence of some iterative schemes is presented. (A brief summary of basic results on convergence of sequences is included as a reminder to the reader.)

As in the first edition, practical justification of the methods is presented through computer examples and exercises. However, a major change from that first edition is that these are now achieved through the use of MATLAB[1] rather than the BASIC programming language. MATLAB is an extremely powerful package for numerical

---

[1] MATLAB is a registered trademark of The MathWorks, Inc.

computing and programming which has many 'add-on' toolboxes. However we use nothing outside the basic MATLAB package in this text. Each chapter concludes with a section describing some of the relevant MATLAB functions – which will often include implementations of methods beyond the scope of this book. It is a mark of the power and success of packages such as MATLAB that programming languages such as FORTRAN which were formerly regarded as 'high-level' languages have become 'low-level' ones in the modern era of scientific computing. The book has an Appendix devoted to the basics of the MATLAB package, its language and programming.

The MATLAB example code used in this book is not intended to exemplify sophisticated or robust pieces of software; it is simply illustrative of the methods under discussion. This is important for a beginning student since writing robust code necessitates taking into account all sorts of difficulties well beyond our present scope. Our objective here is to get the basic ideas across. The interested student can ponder some of these difficulties in subsequent courses.

As to the content of the book: Chapter 1 is devoted to floating-point arithmetic and errors and, throughout the book we take time to consider the precision of our numerical solutions, and how this can be safeguarded or improved. Subsequent chapters deal with the *solution of* (nonlinear) *equations*, now including pairs of equations in two unknowns (Chapter 2); the *approximate evaluation of functions*, both by series approximations and the CORDIC algorithms that are used in almost all calculators and PCs (Chapter 3); *interpolation* with both polynomial and cubic spline interpolation discussed (Chapter 4); *numerical calculus* including integration, differentiation and location of extrema (Chapter 5); *ordinary differential equations* including Runge-Kutta and multistep methods as well as systems of differential equations and shooting and finite difference methods for boundary value problems (Chapter 6). Chapter 7 is devoted to *linear equations*. This fundamental topic is deliberately left to the end to allow the earlier chapters to motivate its treatment. This chapter has now been augmented with a section on eigenvalues. However, Chapter 7 is independent of any of the specific material and so can easily be covered earlier if desired.

The intention of this book remains to provide the student with an introduction to this subject which is not, in its combined demands of computing, motivation, manipulation and analysis, paced such that only the most able can 'see the wood for the trees'. The major effort of programming is therefore removed from the reader, as are the harder parts of the analysis. The algebraic manipulation is largely unavoidable but is carefully explained – especially in the early stages. The motivation for the numerical methods is provided by the examples – which are not all of types that can be solved directly – and the numerical methods themselves provide the motivation for the necessary analysis.

In writing the Preface for the first edition, it was important to thank two good friends David Towers, the series editor, and Charles Clenshaw who was a great source of help in the writing process. To those I should add many colleagues who have influenced me over the last ten years. Where the influence is positive it is to their credit, if it is ever negative it is probably because I did not take their good advice! Notable among these are Frank Olver, Dan Lozier, Alan Feldstein, Jim Buchanan, George Nakos and Bob Williams. The struggles of hundreds of students have also helped me see the difficulties, and, I hope, have led to better explanations of many topics.

Annapolis   PETER R. TURNER

1 Number Representations and Errors

Aims and objectives

In this chapter, we study in broad outline the floating point representation used in computers for real numbers and the errors that result from the finite nature of this representation. Our objective is not to impart a detailed understanding of this large topic, but rather to give a good general understanding of how the computer represents and manipulates numbers. We see later that such considerations affect the choice and design of computer algorithms for solving higher-order problems.

1.1   Introduction

In this book we shall consider some of the fundamental ideas of mathematical – and, in particular, numerical – computing. These in turn form the basis of most of the techniques of scientific computing, which are widely used in almost every branch of scientific work, and in business and economic applications.

However, we do not deal here solely with the methodology of solving these problems on a computer. We shall also consider in some detail the underlying mathematical ideas – the reasons why our methods work (or do not work!). This text therefore serves not just as an introduction to this area of applicable mathematics, *numerical analysis* or *scientific computing*, but also to some of the ideas of mathematical analysis. Scientific computing nowadays really includes more than just numerical methods and their analysis. Modelling and symbolic computing are both aspects of this rapidly changing and growing discipline, but we concentrate here on the numerical aspects.

After a brief look at the ways in which numbers are stored and manipulated in computers, and the errors which result from this representation and arithmetic, we shall be concerned with fundamental mathematical problems such as the solution of equations, the evaluation of functions and their integrals, and the solution of differential equations. Your previous experience may suggest that these are not really problems at all but, as we shall see in these various contexts, the range of problems which can be solved *exactly* is very limited indeed. We therefore require efficient methods of obtaining good *approximate* solutions and (and this is one of the major aspects of numerical analysis) the ability to estimate how good, or bad, our approximations are.

Before proceeding to any of the specific topics of this introductory chapter, it is worth pointing out that we shall be introducing several definitions and concepts which may be unfamiliar. The reader should not spend time trying to master all these immediately but should rather try to acquire a rough idea of the sorts of difficulties which can arise from the computer solution of mathematical problems.

1.2 Floating-point numbers

Within any electronic computer, since the machine itself is finite, we can represent only a finite set of numbers. But of course the set of real numbers which we use in, for example, calculus is infinite. It is therefore necessary to decide on an efficient method of representing numbers in a computer so that we reach a generally acceptable compromise between the *range* of numbers available and the *accuracy* or *precision* of their representation. The way this is usually achieved is to use the *floating-point representation* which is, in principle, much like the scientific notation used for many physical constants and in the display of hand-calculators.

Example 1   Floating-point representations of π

The constant $\pi = 3.1416$ to 4 decimal places, or 5 significant figures, would be represented in a normalized decimal floating-point system using 4 significant figures by

$$\pi \approx +3.142 \times 10^0$$

With 6 significant figures, we would have

$$\pi \approx +3.14159 \times 10^0$$

This representation consists of three pieces: the *sign*, the *exponent* (in this case 0) and the *mantissa*, or *fraction*, 3.14 . . . .

In the *binary* floating-point system using 18 significant binary digits, or *bits*, we would have

$$\pi \approx +1.10010010000111111 \times 2^1$$

Here the digits following the *binary point* represent increasing powers of $1/2$. Thus the first 5 bits represent

$$\left(1 + \frac{1}{2} + \frac{0}{4} + \frac{0}{8} + \frac{1}{16}\right) \times 2^1 = 3.125$$

The decimal base is commonly used for calculators while the use of binary floating-point is almost universal for computers. Some computer systems use the *hexadecimal* (base 16) system. This is easily obtained from the binary representation by grouping bits in groups of four and adjusting the exponent to be a multiple of 4. (Note that $16 = 2^4$.) We shall concentrate on the two common systems.

In general, a positive number x can be expressed in terms of an arbitrary base β as

$$x = (a_n a_{n-1} \ldots a_1 a_0 . a_{-1} a_{-2} \ldots a_{-m})_\beta$$

which represents the quantity

$$a_n \beta^n + a_{n-1} \beta^{n-1} + \cdots + a_1 \beta^1 + a_0 \beta^0 + a_{-1} \beta^{-1} + a_{-2} \beta^{-2} + \cdots + a_{-m} \beta^m$$

and each of the base-β digits $a_n, a_{n-1}, \ldots, a_1, a_0, a_{-1}, a_{-2}, \ldots, a_{-m}$ is an integer in the range $0, 1, 2, \ldots, \beta - 1$. The representation above could be rewritten using summation notation as

$$x = \sum_{k=-m}^{n} a_k \beta^k$$

Such a number may be expressed in (base-β) floating-point form as

$$x = f \times \beta^E \tag{1.1}$$

where the mantissa f satisfies

$$1 \leq f < \beta \tag{1.2}$$

This last condition (1.2) is the basis of the *normalized* floating-point system used in almost all computer systems. Equation (1.1) is replaced by an approximate equation in which the mantissa is represented to a fixed number of significant figures. The mantissa is therefore of the form

$$
\begin{aligned}
f &= b_0 + b_1 \beta^{-1} + b_2 \beta^{-2} + \cdots + b_N \beta^N \\
&= \sum_{k=0}^{N} b_k \beta^{-k}
\end{aligned}
\tag{1.3}
$$

An unnormalized system need not satisfy (1.2). Such unnormalized numbers are used in computer systems to extend the range of representable numbers in special situations. We are not interested here in the fine detail of the floating-point system, only its basic properties.

We note that in the *binary* system $\beta = 2$, (1.2) reduces to $1 \leq f < 2$. We deduce that

$$b_0 = 1 \tag{1.4}$$

in the binary representation (1.3) of the mantissa. This particular bit is not stored explicitly in normalized binary floating-point representations. It is referred to as the *implicit* bit. (Of course, it must be used in any arithmetic processing of the data.)

The most commonly used floating-point representations are the IEEE binary floating-point standards. There are two such formats: *single* and *double precision*. IEEE single precision uses a 32-bit *word* to represent the sign, exponent and mantissa. Double precision uses 64 bits. These bits are distributed as shown in Table 1.1.

In Example 1, there was an implicit assumption that the representation uses *symmetric rounding* so that a number is represented by the nearest member of the set of representable numbers. This is by no means the universal *rounding mode* used by computer systems. The IEEE standard requires the inclusion of symmetric rounding, *rounding towards zero* (or *chopping*), and the directed rounding modes towards either $+\infty$ or $-\infty$.

Using chopping in the decimal floating-point approximations of π given in Example 1, we would obtain $\pi \approx +3.141 \times 10^0$ and $\pi \approx +3.14159 \times 10^0$. With 5 significant figures, chopping would yield $\pi \approx +3.1415 \times 10^0$.

We shall, for the most part, assume symmetric rounding. This is not quite as simple as described here because of the ambiguity when a number falls exactly midway between two representable numbers. The details of how such rounding decisions are made are not important to our present work.

It should be noted here that most computer systems and programming languages allow quantities which are known to be integers to be represented in their exact binary form. This necessarily restricts the size of integers which can be stored in a single computer *word* (that is, some fixed number of bits) but avoids the introduction of any further errors as the result of the representation.

The errors in the floating-point representation of real numbers also affect subsequent calculations using this data. We shall consider this in some detail in the next few sections but, for now, we illustrate the possible effects with some examples.

Table 1.1   IEEE normalized binary floating-point representations

Numbers of bits	*Single-precision*	*Double-precision*
Sign	1	1
Exponent	8	11
Mantissa (including implicit bit)	24	53

Example 2   Operations on a hypothetical decimal computer

Consider the following simple operations on a hypothetical decimal computer in which the result of every operation is rounded to 4 significant digits.

1   The addition

$$1.234 \times 10^0 + 1.234 \times 10^{-1} = 1.3574 \times 10^0$$
$$\approx 1.357 \times 10^0$$

has a rounding error of 4×10^{-4} which is also true of the corresponding subtraction

$$1.234 \times 10^0 - 1.234 \times 10^{-1} = 1.1106 \times 10^0$$
$$\approx 1.111 \times 10^0$$

Multiplication of the same pair of numbers has the exact result 1.522756×10^{-1} which would be rounded to 1.523×10^{-1}. Again there is a small rounding error.

2   The somewhat more complicated piece of arithmetic

$$\frac{1.234}{0.1234} - \frac{1.234}{0.1233}$$

demonstrates some of the pitfalls more dramatically. Proceeding in the order suggested by the layout of the formula, we obtain

$$\frac{1.234}{0.1234} \approx 1.000 \times 10^1, \text{ and}$$
$$\frac{1.234}{0.1233} \approx 1.001 \times 10^1$$

from which we get the result -0.001×10^1 which becomes -1.000×10^{-2} after normalization.

3   If we perform the last calculation rather differently, we can first compute

$$\frac{1}{0.1234} - \frac{1}{0.1233} \approx 8.104 - 8.110 = -6.000 \times 10^{-3}$$

Multiplying this result by 1.234 we get -7.404×10^{-3} which is much closer to the correct result which, rounded to the same precision, is 8.110×10^{-3}.

The examples above illustrate some basic truths about computer arithmetic and rounding errors – but again it is worth emphasizing that you should not be perturbed if all the details are not immediately clear. First, we see that, because the number representation is not exact, the introduction of further error as a result of arithmetic is inevitable. In particular the subtraction of 2 numbers of similar magnitudes results invariably in the loss of some precision. (The result in part 2 of Example 2 has only *1* significant figure since the zeros introduced after the normalization are entirely spurious.) There is, in general, nothing that can be done to avoid this phenomenon but, as part 3 demonstrates, we can alleviate matters by taking care over the order in which a particular computation takes place.

Two related problems can arise from the finiteness of the set of machine numbers. Suppose that, on the same hypothetical machine as we used in Example 2, there is just 1 (decimal) digit allocated to the exponent of a floating-point number. Then, since

$$\left(3.456 \times 10^3\right) \times \left(3.456 \times 10^7\right) = 3456 \times 34560000$$
$$= 119439360000$$
$$\approx 1.194 \times 10^{11}$$

the result of this operation is too large to be represented in our hypothetical machine. Such an event is called *overflow*. Similarly, if the (absolute value of a) result is too small to be represented, it is termed *underflow*. For our hypothetical computer this would occur if the exponent of the normalized result were less than -9.

Fortunately such catastrophic situations quite rarely result from correct computation, but even some apparently straightforward tasks must be programmed very carefully in order to avoid overflow or underflow since their occurrence will cause a program to fail. Many systems will not fail on underflow but will simply set such results to zero. But that can often result in a later failure, or in meaningless answers, so care should still be exercised.

We have already pointed out that the floating-point representation is not the only one used in computers since integers can be stored exactly, or in *fixed-point form*. For several years, in the infancy of the computer age, this was the only representation available and calculations needed to be scaled to avoid overflow and underflow. Today, a similar deficiency is becoming apparent in the floating-point systems and new number representations have been proposed in order to try to avoid overflow and underflow altogether.

1.2.1 MATLAB number representation

In all essential respects, MATLAB uses only one type of number – IEEE double precision floating-point. Strictly speaking, MATLAB uses pairs of these to represent double precision floating point *complex* numbers, but that will not affect much of what we do here. Integers are stored in MATLAB as 'floating integers'

Table 1.2   Key constants of MATLAB's arithmetic

MATLAB variable	Meaning	Value
eps	Machine unit	$2^{-52} \approx 2.22044604925031e-016$
realmin	Smallest positive number	$2.2250738585072e-308$
realmax	Largest positive number	$1.79769313486232e+308$

which means, in essence, that integers are stored in their floating-point representation.

From Table 1.1 above, we see that this representation uses 11 bits for the binary exponent which therefore ranges from about -2^{10} to $2^{10} = 1024$. (The actual range is not exactly this because of special representations for small numbers and for $\pm\infty$.) The mantissa has 53 bits including the implicit bit. If $x = f \times 2^E$ is a normalized MATLAB floating-point number then $f \in [1, 2)$ is represented by

$$f = 1 + \sum_{k=1}^{52} b_k 2^{-k}$$

Since $2^{10} = 1024 \approx 10^3$, these 53 significant bits are equivalent to approximately 16 significant decimal digits accuracy in the MATLAB representation.

The fact that the mantissa has 52 bits *after the binary point* means that the next machine number greater than 1 is $1 + 2^{-52}$. This gap is called the *machine unit*, or machine *epsilon*. This and other key constants of MATLAB's arithmetic are easily obtained (see Table 1.2).

Here, and throughout, the notation *xen* where x, n are numbers stands for $x \times 10^n$. Thus eps $\approx 2.2 \times 10^{-16}$. The quantity 10^{-8} may be written in MATLAB as 1e–8.

In MATLAB, neither underflow nor overflow cause a 'program' to stop. Underflow is replaced by zero, while overflow is replaced by $\pm\infty$. This allows subsequent instructions to be executed and may permit meaningful results. Frequently, however, it will result in meaningless answers such as $\pm\infty$ or NaN, which stands for *Not-a-Number*. NaN is the result of indeterminate arithmetic operations such as $0/0$, ∞/∞, $0 \cdot \infty$, $\infty - \infty$ etc.

Exercises: Section 1.2

1 Express the base of natural logarithms e as a normalized floating-point number, using both chopping and symmetric rounding, for each of the following systems:

(*a*) base 10 with 4 significant figures
(*b*) base 10 with 7 significant figures
(*c*) base 2 with 10 significant bits

2 Write down the normalized binary floating-point representations of $1/3$, $1/5$, $1/6$, $1/7$, $1/9$ and $1/10$. Use enough bits in the mantissa to see the recurring patterns.

1.3   Sources of errors

Of course, the most common source of error in any mathematical process is just human blunder, but for the purpose of this discussion we shall assume infallibility of both the mathematician and the computer programmer. Even in this fantasy world, errors are an everyday fact of computational life. The full analysis of these errors – their sources, interactions and combined effect – and the right strategy to minimize their effect on a complicated computational process can be a very difficult task, and is certainly way beyond the scope of this book.

In this section, we consider just three of the principal sources of error. In several places we shall cite examples of methods which will probably be unfamiliar at this stage. They are purely illustrative and you should not be too concerned about following the details.

1.3.1   Rounding errors

We have already seen that these arise from the storage of numbers to a fixed number of binary or decimal places or significant figures.

Example 3   Effect of rounding error

The equation

$$x^2 - 10.1x + 1 = 0$$

has the exact solutions 10 and 0.1 but if the quadratic formula is used on a hypothetical 4-decimal digit machine the roots obtained are 10.00 and 0.1005. The errors are due to rounding of all numbers in the calculation to 4 significant figures.

If, however, we use this same formula for the larger root together with the fact that the product of the roots is 1, then, of course, we obtain the values 10.00 and $1.000/10.00 = 0.1000$. We thus see that the effect of the rounding error can be reduced (in this case eliminated) by the careful choice of numerical process, or the *design of the algorithm*. (It is worth noting that a quadratic equation with real roots should *always* be solved this way.)

The results in Example 3 are based on symmetric rounding. If chopping were used throughout, the larger root would be obtained as 9.995. The smaller would then be $1/9.995 = 0.100\,05$ which with chopping gives 0.1000, as expected.

1.3.2 Truncation error

Truncation error is the name given to the errors which result from the many ways in which a numerical process can be cut short or truncated. Obvious examples are the use of a finite number of terms of a series to estimate its sum – for example, in routines for evaluating elementary functions (see Chapter 3). With insufficient care, such errors can lead us back to blunders since, for example, working to any specified accuracy, the harmonic series

$$1 + \frac{1}{2} + \frac{1}{3} + \cdots = \sum_{n=1}^{\infty} \frac{1}{n}$$

appears to converge. If we work to 2 decimal places with symmetric rounding, then every term beyond $n = 200$ will be treated as zero. We might therefore conclude that, to 2 decimal places,

$$\sum_{n=1}^{\infty} \frac{1}{n} = \sum_{n=1}^{200} \frac{1}{n} = 5.88$$

But we know this sum is in fact *infinite*!

The difficulty here is caused partly by truncation error and partly by rounding error. In order to claim that some finite number of terms gives the sum of an infinite series to a specified accuracy, we must ensure that the sum of all the remaining terms in the series has no effect to that accuracy. (It would also be necessary for such a routine to work to greater precision than is required in the final result.)

Example 4   Determine the number of terms of the exponential series required to estimate e to three decimal places

We consider here just the truncation error by assuming that the calculations are performed exactly. Now

$$e = \exp(1) = 1 + 1 + \frac{1}{2!} + \frac{1}{3!} + \cdots = \sum_{n=0}^{\infty} \frac{1}{n!}$$

The tail of this series can easily be bounded for any positive integer N by a geometric series as follows:

$$\sum_{n=N}^{\infty} \frac{1}{n!} = \frac{1}{N!} + \frac{1}{(N+1)!} + \cdots$$

$$< \frac{1}{N!} \left(1 + \frac{1}{N} + \frac{1}{N^2} + \cdots \right)$$

$$= \frac{1}{N!} \cdot \frac{N}{N-1} = \frac{1}{(N-1)(N-1)!}$$

Now, $1/(5 \times 5!) = 1.6667 \times 10^{-3}$ while $1/(6 \times 6!) = 2.3148 \times 10^{-4}$. Thus we may estimate e to the required accuracy by

$$e \approx \sum_{n=0}^{6} \frac{1}{n!} = 2.7181$$

Furthermore, we know this is an *underestimate* since all the terms are positive. We may thus deduce that

$$2.7181 = \sum_{n=0}^{6} \frac{1}{n!}$$

$$\leq e$$

$$= \sum_{n=0}^{6} \frac{1}{n!} + \sum_{n=7}^{\infty} \frac{1}{n!}$$

$$\leq 2.7181 + 2.3148 \times 10^{-4}$$

$$\approx 2.7183$$

Obviously this calculation is also subject to rounding errors, but, in this case, the accumulated rounding error has been controlled by performing the calculations to much greater precision than the 4 decimals required in the result.

The errors in approximating a function by the first N terms of its Taylor series, or in approximating an integral by Simpson's rule, are examples of truncation error, as is the error incurred by stopping an iteration when two iterates agree to a certain accuracy.

1.3.3 Modelling errors and ill-conditioning

Frequently in the mathematical modelling of physical or economic situations, simplifying assumptions are made which inevitably result in output errors. For example, in the equations of motion for a simple pendulum, we often make the simplification $\sin \theta = \theta$ for small amplitude oscillations. Such assumptions must be analyzed to estimate their effect on the results. Sometimes this might be achieved by, for example, perturbing the value of some assumed constant and measuring the consequent change in the results.

If the model and the numerical method are stable, then small perturbations in the input should yield small changes in the output. This is very closely related to the notion of conditioning. We illustrate the idea of an *ill-conditioned* problem with a classical example due to Wilkinson of finding the roots of the polynomial equation

$$p(x) = (x-1)(x-2)\cdots(x-20) = 0$$

where p is given in the form

$$p(x) = x^{20} + a_{19}x^{19} + \cdots + a_1 x + a_0$$

If the coefficient $a_{19} = -210$ is changed by about 2^{-22}, then the resulting polynomial has only 10 real zeros, and 5 complex conjugate pairs. The largest real root is now at $x \approx 20.85$. A change of only 1 part in about 2 billion in just one coefficient has certainly had a significant effect!

Very rarely does a physical situation have such inherent instability and so, if such ill-conditioning occurs in practice, the fault probably lies in the mathematical model and/or the data, or in the mathematical techniques being used. All these should therefore be re-examined. The science of Chaos has developed from an attempt to understand genuine instabilities of this form where small changes in the inputs can result in large and unpredictable responses. One of the key questions in determining whether a system is chaotic is to show that the 'chaotic' behaviour is not just the result of sensitivity to rounding errors in the initial conditions or in the numerical computation.

Exercises:
Section 1.3

1 Perform the calculations of Example 3 (p. 8). Repeat the same computation using chopping on a hypothetical 6-decimal digit machine.

2 Write a computer (MATLAB) program to 'sum' the harmonic series $1 + \frac{1}{2} + \frac{1}{3} + \cdots$ stopping when

    (*a*)   all subsequent terms are zero to 3 decimal places, and

    (*b*)   two successive 'sums' are equal to 6 decimal places.

    (Extension) Try to obtain the 'best approximation of ∞' you can by summing terms of the harmonic series.

3 How many terms of the series expansion

$$\cosh x = 1 + \frac{x^2}{2!} + \frac{x^4}{41} + \cdots = \sum_{k=0}^{\infty} \frac{x^{2k}}{(2k)!}$$

are needed to estimate $\cosh(1/2)$ with a truncation error less than 10^{-8}? Check that your answer achieves the desired accuracy by using MATLAB to sum the terms and comparing the result with the built-in cosh function.

4 For what range of values of x will the truncation error in the approximation

$$\exp x = e^x \approx \sum_{k=0}^{6} \frac{x^k}{k!}$$

be bounded by 10^{-10}?

1.4   Measures of error and precision

The two most common error measures are the absolute and relative error in approximating a number x by a (nearby) number $\widehat{x}$.

The *absolute error* is defined to be $|x - \widehat{x}|$; this corresponds to the idea that x and $\hat{x}$ agree to a number of decimal places.

The *relative error* is usually defined to be $|x - \widehat{x}|/|x| = |1 - \widehat{x}/x|$ which corresponds to agreement to a number of significant figures. Because we typically do not have the true value of x available, it is common to replace the x in the denominator by its approximation $\widehat{x}$. In this case the relative error is given by $|x - \widehat{x}|/|\widehat{x}| = |1 - x/\widehat{x}|$. (The asymmetric nature of this definition is somewhat unsatisfactory, but this can be overcome by the alternative notion of relative precision.)

Example 5   **Find the absolute and relative errors in approximating π by 3.1416. What are the corresponding errors in the approximation $100\pi \approx 314.16$?**

Obviously we cannot find the absolute errors exactly since we do not know a numerical value for π *exactly*. However, we can get a good idea of these errors by using the more accurate approximation

$$\pi \approx 3.14159265358979$$

obtained from MATLAB. Then the absolute error can be estimated as

$$|\pi - \hat{\pi}| = |3.14159265358979 - 3.1416| = 7.346 \times 10^{-6}$$

The corresponding relative error is

$$\frac{|\pi - \hat{\pi}|}{|\pi|} = \frac{|3.14159265358979 - 3.1416|}{3.14159265358979}$$
$$= 2.3383 \times 10^{-6}$$

Now using the approximations to 100π, we get the absolute error

$$|314.159265358979 - 314.16| = 7.346 \times 10^{-4}$$

which is exactly 100 times the error in approximating π, of course. The relative error is, however, unchanged since the factor of 100 affects numerator and denominator the same way:

$$\frac{|314.159265358979 - 314.16|}{314.159265358979} = \frac{|3.14159265358979 - 3.1416|}{3.14159265358979}$$
$$= 2.3383 \times 10^{-6}$$

As a general principle, we expect absolute error to be more appropriate for quantities close to unity, while relative error seems more natural for large numbers or those close to zero. However, we must be wary – a 1% relative error in the distance travelled by a Moon shot could easily be sufficient to miss the Moon altogether!

For approximations to functions (as opposed to their values at particular points), other measures are used. These *metrics* are usually expressed in the form of measures of absolute error but are readily adapted for the purpose of relative error. The three most commonly used are defined for approximating a function f on an interval $[a, b]$ by another function (or polynomial) p:

- The *supremum* or *uniform* or L_∞ *metric*

$$||f - p||_\infty = \max_{a \leq x \leq b} |f(x) - p(x)|$$

- The L_1 *metric*

$$||f - p||_1 = \int_a^b |f(x) - p(x)| dx$$

- The L_2 or *least-squares metric*

$$||f - p||_2 = \sqrt{\int_a^b |f(x) - p(x)|^2 dx}$$

These metrics, or *norms*, are examples of a quite sophisticated mathematical notion of measurement of distance and will be totally unfamiliar at this stage. All that is required here is a very superficial understanding of their meaning and purpose.

The first measures the extreme discrepancy between f and the approximating function p while the others are both measures of the 'total discrepancy' over the interval.

Example 6   Find the L_∞, L_1, and L_2 errors in approximating $\sin\theta$ by θ over the interval $[0, \pi/4]$

We begin with a plot of the function and its approximant:

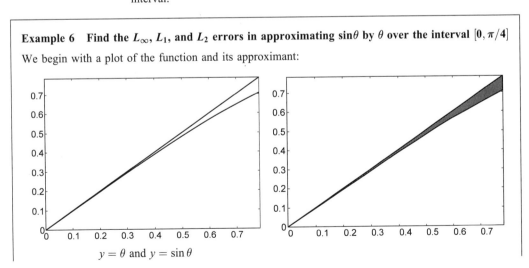

$y = \theta$ and $y = \sin \theta$

For the L_∞ error, we need to find the maximum discrepancy between the two curves. From the graphs, or from simple calculus, this clearly occurs at $\theta = \pi/4$ so that

$$\|\sin\theta - \theta\|_\infty = \pi/4 - \sin(\pi/4)$$
$$= 7.829\,1 \times 10^{-2}$$

For the L_1 error we need to compute the area of the shaded region between the curves:

$$\|\sin\theta - \theta\|_1 = \int_0^{\pi/4} (\theta - \sin\theta)d\theta$$
$$= \frac{1}{32}\pi^2 + \frac{1}{2}\sqrt{2} - 1 = 1.553\,2 \times 10^{-2}$$

The L_2 error is given by

$$\|\sin\theta - \theta\|_2^2 = \int_0^{\pi/4} (\theta - \sin\theta)^2 d\theta$$
$$= 6.972\,8 \times 10^{-4}$$

so that $\|\sin\theta - \theta\|_2 = \sqrt{6.972\,8 \times 10^{-4}} = 2.640\,6 \times 10^{-2}$.

The calculus proofs of these results are left as exercises.

Of course, if all we know are the values of f at the discrete set of points $x_0, x_1,$ x_N, then the corresponding discrete measures must be used:

$$\max_{k=0,1,\ldots,N} |f(x_k) - p(x_k)|$$

$$\sum_{k=0}^N |f(x_k) - p(x_k)|, \text{ and}$$

$$\sqrt{\sum_{k=0}^N |f(x_k) - p(x_k)|^2}$$

The last of these is the measure of total error commonly used in obtaining a least-squares fit to experimental data.

In these cases where we have data at just a finite set of points, we can use polynomial interpolation (Chapter 4) to obtain a polynomial p such that $f(x_k) = p(x_k)$ for each k. This would make each of these measures of error zero. However this may not be as good an idea as it first appears. First, the data may be subject to experimental errors so that fitting it *exactly* is merely perpetuating these errors; secondly, information on the form of f may tell us from which class of functions to choose p; thirdly, if there are very many data points we do not wish to compute a very high-degree polynomial in full.

Exercises: **Section 1.4**	**1**	Find the absolute and relative errors in the decimal approximations of e found in Exercise 1, Section 1.2 (p. 7).

1 Find the absolute and relative errors in the decimal approximations of e found in Exercise 1, Section 1.2 (p. 7).

2 What is the absolute error in approximating $1/3$ by 0.3333? What is the corresponding relative error?

3 Find the absolute and relative errors in approximating $1/3$, $1/5$, $1/6$, $1/7$, $1/9$ and $1/10$ by their binary floating-point representations using 12-bit mantissas (see Exercise 2, Section 1.2, p. 7).

4 Suppose the function e^x is approximated by $1 + x$ on $[0, 1]$. Find the L_∞, L_1 and L_2 measures of the error in this approximation.

1.5 Floating-point arithmetic

In this section, we discuss briefly the error analysis of floating-point arithmetic and some of its implications for computer design. For completeness, we shall present the general results for an arbitrary base β – typically 2 for computers and 10 for calculators.

We recall from Section 1.2 that a positive number x will be represented in normalized floating-point form by

$$\widehat{x} = \widehat{f}\beta^E \tag{1.5}$$

where the fraction $\widehat{f}$ consists of a fixed number, $(N + 1)$ say, of base-β digits:

$$\widehat{f} = \widehat{d_0}.\widehat{d_1}\widehat{d_2}\ldots\widehat{d_N} \tag{1.6}$$

where $1 \le \widehat{d_0} < \beta$ and $0 \le \widehat{d_k} < \beta$ for $k = 1, 2, \ldots, N$.

The absolute error in representing $x = f\beta^E = \beta^E(d_0.d_1d_2\ldots)$ by $\widehat{x}$ is therefore

$$|x - \widehat{x}| = \beta^E \left| f - \widehat{f} \right| \tag{1.7}$$

If $\widehat{f}$ is obtained from f by chopping then $\widehat{d_k} = d_k$ for $k = 0, 1, \ldots, N$ so that (1.7) becomes

$$|x - \widehat{x}| = \beta^E \sum_{k=N+1}^{\infty} d_k\beta^{-k} \le \beta^{E-N}$$

Symmetric rounding is equivalent to first adding $\beta/2$ to d_{N+1} and then chopping the result. It follows that, for symmetric rounding,

$$|x - \widehat{x}| \le \frac{\beta^{E-N}}{2} \tag{1.8}$$

Of more interest for floating-point computation is the size of the relative error:

$$\frac{|x - \widehat{x}|}{|x|} = \frac{\left|f - \widehat{f}\right|}{|f|}$$

In the same manner as above, we find that

$$\frac{\left|f - \widehat{f}\right|}{|f|} \leq \begin{cases} \dfrac{1}{\beta^N} & \text{for chopping} \\ \dfrac{1}{2\beta^N} & \text{for symmetric rounding} \end{cases} \tag{1.9}$$

To prove the 'chopping' result, note that

$$\frac{\left|f - \widehat{f}\right|}{|f|} = \frac{\sum_{k=N+1}^{\infty} d_k \beta^{-k}}{|f|} \leq \frac{\beta^{-N}}{|f|} \leq \beta^{-N}$$

since $|f| \geq 1$ and $d_k \leq \beta - 1$ for each k. The 'rounding' result is proved similarly.

Thus far we have considered only the floating-point representation errors. More important to the error analysis of any computation is the propagation of these errors through the various arithmetic operations. One of the requirements of the IEEE floating-point standards is that the stored result of any floating-point operation equals the correctly rounded value of the *exact* result, assuming the data were themselves exact.

This requirement implies that the final rounding and normalization result in a relative error no greater than the bounds in (1.9). However this is not the only source of error since the data are usually not exact. If ◆ in $x◆y$ stands for one of the standard arithmetic operations $+, -, \times, \div$ then to analyze floating-point arithmetic errors fully we require bounds on the quantity

$$\frac{\left|\widehat{(\widehat{x}◆\widehat{y})} - (x◆y)\right|}{|(x◆y)|}$$

Such bounds vary according to the nature of the operation ◆. Analyzing a full computational process with this level of detail is rarely necessary, and is certainly beyond our current aims.

In the remainder of this section we shall use some examples to illustrate some of the behaviour of floating-point arithmetic.

Example 7   Subtraction of nearly equal numbers can result in large relative rounding errors

Let $x = 0.12344$, $y = 0.12351$ and suppose we compute the difference $x - y$ using a 4-decimal digit floating-point computer. Then

$$\widehat{x} = (1.234)10^{-1}, \quad \widehat{y} = (1.235)10^{-1}$$

so that

$$(\widehat{\hat{x} - \hat{y}}) = \hat{x} - \hat{y} = -(0.001)10^{-1}$$
$$= -(1.000)10^{-4}$$

while $x - y = -(0.7)10^{-4}$. The relative error is therefore

$$\frac{(1.0 - 0.7)10^{-4}}{(0.7)10^{-4}} = \frac{3}{7} = 0.428\,57$$

That is, there is an error of approximately 43%.

The next example, in terms of absolute errors, illustrates a common drawback of simplistic error bounds: they are often unrealistically pessimistic.

Example 8   Using 'exact' arithmetic

$$\frac{1}{11} + \frac{1}{12} + \cdots + \frac{1}{20} = 0.668771$$

to 6 decimal places; however, rounding each term to 4 decimal places we obtain the approximate sum 0.6687

The accumulated rounding error is only about 7×10^{-5}. The error for each term is bounded by 5×10^{-5}, so that the cumulative error bound for the sum is 5×10^{-4}, which is about 7 times the actual error committed. Error bounds can often be much worse than the actual error.

This phenomenon is by no means unusual and leads to the study of probabilistic error analyses for floating-point calculations. For such analyses to be reliable, it is important to have a good model for the distribution of numbers as they arise in scientific computing. It is a well established fact that the fractions of floating-point numbers are logarithmically distributed. One immediate implication of this distribution is the rather surprising statement that the proportion of (base-β) floating-point numbers with leading digit n is given by

$$\log_\beta \left(\frac{n+1}{n} \right) = \frac{\log(n+1) - \log n}{\log \beta}$$

In particular, 30% of decimal numbers have leading significant digit 1 while only about 4.6% have leading digit 9.

Example 9   The base of your computer's arithmetic is 2

Let n be any positive integer and let $x = 1/n$. Then it is easy to check that $(n + 1)x - 1 = x$ so that the MATLAB command

```
x = (n + 1) * x - 1;
```

should leave the value of x unchanged however many times we repeat it. In the table below we show the effect of doing this 10 and 30 times for each $n = 1, 2, \ldots, 10$. The code used to generate one of these columns was:

```
» for n=1:10
    x=1/n;
    for k=1:10
        x=(n+1)*x-1;
    end
    a(n)=x;
end
```

n	$x = 1/n$	k=1:10	k=1:10, format long	k=1:30
1	1.00000000000000	1.0000	1.00000000000000	1
2	0.50000000000000	0.5000	0.50000000000000	0.5
3	0.33333333333333	0.3333	0.33333333331393	−21
4	0.25000000000000	0.2500	0.25000000000000	0.25
5	0.20000000000000	0.2000	0.20000000179016	6.5451e006
6	0.16666666666667	0.1667	0.16666666069313	−4.7664e+008
7	0.14285714285714	0.1429	0.14285713434219	−9.8171e+009
8	0.12500000000000	0.1250	0.12500000000000	0.125
9	0.11111111111111	0.1111	0.11111116045436	4.9343e+012
10	0.10000000000000	0.1000	0.10000020942783	1.4089e+014

Initially the results of 10 iterations appear to conform to our expectations that x remains unchanged. However the fourth column shows that several of these values are already contaminated with errors. By the time 30 iterations are performed, many of the answers are not recognizable as being $1/n$.

On closer inspection, we see that the ones which remain fixed are those where n is a power of 2. This is because such numbers are represented *exactly* in the binary floating-point system. For all the others there is some initially small representation error: it is the propagation of this error through this computation which results in the enormous errors here.

For example with $n = 3$, the initial value $\widehat{x}$ is not exactly 1/3 but is instead

$$\widehat{x} = \frac{1}{3} + \delta$$

where δ is the rounding error in the floating-point representation of $1/3$.

Now the operation

x=(n+1)*x-1;

has the effect of magnifying this error. After one iteration, we have (ignoring the final rounding error)

$$x = 4\widehat{x} - 1 = 4\left(\frac{1}{3} + \delta\right) - 1 = \frac{1}{3} + 4\delta$$

which shows that the error is increasing by a factor of 4 with each iteration.

In the particular case of $1/3$, it can be shown that the initial representation error in IEEE double precision floating-point arithmetic is approximately $-\frac{1}{3}2^{-54}$. Multiplying this by $4^{30} = 2^{60}$ yields an estimate of the final error equal to $-\frac{1}{3}2^{6} = -21.333$ which explains the entry -21 in the final column for $n = 3$.

Try repeating this experiment to verify that your calculator uses base 10 for its arithmetic.

This chapter has provided a very superficial overview of an extensive subject, and has, I hope, raised at least as many questions in your mind as it has answered.

Exercises: Section 1.5

1 Let $x = 1.3576$, $y = 1.3574$. For a hypothetical four decimal digit machine, write down the representations $\widehat{x}, \widehat{y}$ of x, y and find the relative errors in the stored results of $x + y, x - y, xy$, and x/y using

   (*a*) chopping, and
   (*b*) symmetric rounding.

2 Try to convince yourself of the validity of the statement that floating-point numbers are logarithmically distributed using the following experiment:

   (*a*) Write MATLAB code which finds the leading significant (decimal) digit of a number in $[0, 1)$. (**Hint**: keep multiplying the number by 10 until it is in $[1, 10)$ and then use the floor function to find its integer part.)
   (*b*) Use the random number generator rand to obtain vectors of uniformly distributed numbers in $[0, 1)$. Form 1000 products of pairs of such numbers and find how many of these have leading significant digit $1, 2, \ldots, 9$.
   (*c*) Repeat this with products of three, four and five such factors.

3 Repeat the previous exercise but find the leading *hexadecimal* (base 16) digits. See how closely these conform to the logarithmic distribution.

4 Consider Example 9 (p. 18). Try to explain the entry in the final column for $n = 5$. Use an argument similar to that used for $n = 3$. (You will find it easier if you just *bound* the initial representation error rather than trying to estimate it accurately.)

2 Iterative Solution of Equations

Aims and objectives

One of the most fundamental mathematical problems is the solution of equations. We begin with a single equation in a single unknown. Most such equations cannot be solved algebraically, and so we rely on *iterative* methods, in which we generate a sequence of approximations that (we hope) converges to the desired solution. The chapter concludes with a section on *systems* of 2 equations in 2 unknowns. In all cases, we attempt to give a basic understanding of both the theoretical development of the methods and their computer implementation.

2.1 Introduction

In this chapter, we are concerned with the problem of solving an equation of the form

$$f(x) = 0 \tag{2.1}$$

The basic principle of all the methods we discuss is to develop a *sequence* of estimates of the required solution which we hope *converges* to this solution. It is therefore necessary to have some understanding of what is meant by these terms. A detailed treatment of this subject is not undertaken here but a brief listing of the basic facts is included at the end of this introductory section for reference purposes. Fuller treatment of the convergence of sequences can be found in standard texts on calculus or elementary real analysis.

The principle of *iterative methods* is that an initial estimate, or guess, of the solution is made and this is continually refined (usually according to some simple rule) in such a way that the subsequent estimates, or *iterates*, get steadily closer to the required solution of the original problem. Because of the repetitious nature of the calculations, such methods are well suited to computer implementation. Some sample MATLAB programs are included. It is worth commenting here that these are not intended to be robust general-purpose programs, but simply to illustrate the methods described. MATLAB includes m-files for some more robust code; we shall discuss these briefly at the end of the chapter.

We shall also discuss the convergence properties of the methods. It is important to determine when a particular method will work – and when it will fail! It is also important to get an idea of how fast a particular iterative sequence converges. The primary motivation for these theoretical aspects is to avoid performing many unnecessary (or fruitless) iterations and to have some idea of the level of precision in a computed answer.

2.1.1 Summary of convergence of sequences

Definition 1 A sequence of real numbers (a_n) converges to a limit L, written

$$a_n \to L \text{ as } n \to \infty$$

or

$$\lim_{n \to \infty} a_n = L$$

if for every (small) $\varepsilon > 0$ there exists a number N such that

$$|a_n - L| < \varepsilon$$

for every $n > N$.

The condition $|a_n - L| < \varepsilon$ can often be usefully rewritten in the form

$$L - \varepsilon < a_n < L + \varepsilon$$

- $1/n^p \to 0$ as $n \to \infty$ for every positive power p
- $x^n \to 0$ as $n \to \infty$ for any $|x| < 1$, (x^n) diverges if $|x| > 1$ or if $x = -1$
- If $a_n \to a$, $b_n \to b$ then
  1. $a_n \pm b_n \to a \pm b$
  2. $a_n b_n \to ab$ (In particular, $ca_n \to ca$ for any constant c.)
  3. $a_n/b_n \to a/b$ provided $b_n, b \neq 0$ (in particular, $1/b_n \to 1/b$)
- **Sandwich Theorem**: If $a_n \leq c_n \leq b_n$ for every n and $\lim a_n = \lim b_n = L$, then $\lim c_n = L$, also.
- If (a_n) is an *increasing* sequence (that is, $a_{n+1} \geq a_n$ for every n) which is *bounded above* (there exists a number $M \geq a_n$ for every n) then (a_n) converges. A similar statement applies to *decreasing* sequences.
- If $a_n \to a$ and f is continuous at a, then $f(a_n) \to f(a)$ (provided the terms of the sequence are in the domain of f).

We shall make use of several of these facts during our study of iterative sequences.

2.2   The bisection method

Perhaps the simplest technique for solving (2.1) is to use the method of *bisection* which is based on a common sense application of the *Intermediate Value Theorem*.

Suppose that f is continuous on an interval $[a, b]$ and that $f(a)f(b) < 0$ (so that the function is negative at one end-point and positive at the other) then, by the intermediate value theorem, there is a solution of (2.1) between a and b.

The basic situation and the first two iterations of the bisection method are illustrated in Figure 2.1. At each stage we set m to be the midpoint of the current interval $[a, b]$. If $f(a)$ and $f(m)$ have opposite signs then the solution must lie between a and m. If they have the same sign then it must lie between m and b. We can therefore replace one of the endpoints by this midpoint and repeat the process on the new smaller interval.

In Figure 2.1, we see that for the original interval $[0, 1.2]$, $m = 0.6$ and $f(a), f(m)$ have the same sign. The endpoint a is replaced by m – which is to say m becomes the new a – and a new m is computed. It is the midpoint of $[0.6, 1.2]$, $m = 0.9$ as shown. This time $f(a)$ and $f(m)$ have opposite signs, $f(a)f(m) < 0$, and so b would be replaced by m.

The process can be continued until the solution has been found to any required accuracy, or *tolerance*. We may safely stop when the length of the interval $b - a$ is smaller than the tolerance.

The bisection method is summarized as follows.

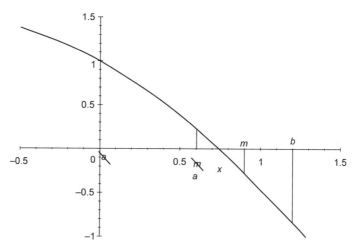

Figure 2.1   The bisection method

Algorithm 1   Bisection method

Input   Equation $f(x) = 0$
            Interval $[a, b]$ such that $f(a)f(b) < 0$
            Tolerance ε
Repeat $m := (a + b)/2$
            if $f(a)f(m) \leq 0$ then $b := m$ else $a := m$
until $b - a < \varepsilon$
Output   Solution lies in the interval $[a, b]$ which has length less than ε

Example 1   Use the bisection method to solve

$$x - \cos x = 0 \qquad\qquad (2.2)$$

with a tolerance of 0.1.

With $f(x) = x - \cos x$, we see easily that

$$f(0) = -1 < 0, \quad f(\pi/2) = \pi/2 > 0$$

Since f is continuous, the intermediate value theorem shows that there is a solution of (2.2) in $[0, \pi/2]$. We set $a = 0$, $b = \pi/2$.

First iteration: $m = \pi/4$
$f(\pi/4) = \pi/4 - \cos \pi/4 = 0.07\,829 > 0$
Hence the solution lies in $[0, \pi/4]$ so we set $b = \pi/4$
Second iteration: $m = \pi/8$
$f(\pi/8) = \pi/8 - \cos \pi/8 = -0.531\,18 < 0$
Hence the solution lies in $[\pi/8, \pi/4]$ so we set $a = \pi/8$
Third iteration: $m = 3\pi/16$
$f(3\pi/16) = 3\pi/16 - \cos 3\pi/16 = -0.242\,42 < 0$
Hence the solution lies in $[3\pi/16, \pi/4]$ so we set $a = 3\pi/16$
Fourth iteration: $m = 7\pi/32$
$f(7\pi/32) = 7\pi/32 - \cos 7\pi/32 = -0.08\,579 < 0$
Hence the solution lies in $[7\pi/32, \pi/4]$ so we set $a = 7\pi/32$

At this point the process stops (because $\pi/4 - 7\pi/32 \approx 0.09\,817\,5 < 0.1$) with output along the lines of

Solution lies in $(7\pi/32, \pi/4)$

or

Solution is $0.736\,31$ with error less than 0.05

Note that $0.736\,31$ is the midpoint of the final interval and so is less than $\varepsilon/2$ away from either end of the interval.

Obviously this would quickly become tedious if we were to try to find this solution to high accuracy by hand using the bisection method. However, the description in Algorithm 1 makes it apparent that the process should be easy to automate. This is illustrated in the next example.

Example 2   Solve (2.2) used in Example 1 using the bisection method in MATLAB. Start with the interval $[0, 1]$ and obtain the solution with an error less than 10^{-5}

First we define a function m-file eq1.m for the equation as follows

```
function y=eq1(x)
y=x-cos(x);
```

Then the following instructions in the MATLAB command window will achieve the desired result:

```
» a=0; b=1;
» tol=1e-5;
» fa=eq1(a); fb=eq1(b);
» while abs(b-a) >2*tol
    m=(a+b)/2; fm=eq1(m);
    if fa*fm<=0, b=m;
    else a=m; end
end
» m=(a+b)/2
m =
    0.739082336425781
```

Note: We could use 2*tol in the test for the while statement because of the final step after the completion of the loop which guarantees that this final solution has the required accuracy. Also we used abs(b-a) to avoid any difficulty if the starting values were reversed.

Of course, the trailing digits in the final answer do not imply that level of precision in the solution – we have only ensured that the true solution is within 10^{-5} of this value.

It should be fairly apparent that the commands used here correspond to those of Algorithm 1. One key difference between that algorithm and the MATLAB commands above is that the latter depend critically on the *name* of the function file whereas in the algorithm the function is one of the inputs. This can be achieved by creating an m-file for the bisection method:

Program	**MATLAB m-file for the bisection method, Algorithm 1**

```
function sol=bisect(fcn,a,b,tol)
% Bisection method for solution of the equation
% fcn(x)=0
% to an accuracy tol, in the interval [a,b]
% The function fcn must be saved as a function m-file
fa=feval(fcn,a); fb=feval(fcn,b);
while abs(b-a)>2*tol
  m=(a+b)/2;
  fm=feval(fcn,m);
  if fa*fm<=0, b=m;
  else a=m; end
end
sol=(a+b)/2;
```

Note: Use of the function feval for 'function evaluation' which we must use to evaluate a function which is passed to an m-file as a parameter.

To see the use of this bisection algorithm function, we resolve the same equation to greater accuracy with the single command:

```
» s=bisect('eq1',0,1,1e-6)
s=
  0.73908519744873
```

Note: The function name 'eq1' must be enclosed in single quotes when used as a parameter for another m-file.

The program above will work provided the function is continuous and provided that the initial values fa and fb have opposite signs. It would be easy to build in a check that this condition is satisfied by inserting the lines

```
if fa*fb>0
  fprintf('Two endpoints have same sign \n')
  return
end
```

immediately before the start of the while loop.

Example 3   **Use the bisection method to find the positive solution of the equation $e^x - 2x - 1 = 0$ by the bisection method**

First we shall try the initial interval $[0.1, 1]$. With this function saved as eq2.m the command

```
» s=bisect('eq2',0.1,1,1e-6)
```

produces the message

Two endpoints have same sign

followed by

Warning: One or more output arguments not assigned during call to 'bisect'.

which is caused by the early exit from the bisection process. It is not an indication of an error.

We can deduce from the intermediate value theorem that the desired solution lies in the interval [1,2]. Using this interval we obtain the solution as follows:

```
» s=bisect('eq2',1,2,1e-6)
s =
    1.25643062591553
```

We conclude this section with an outline of the proof that this method indeed converges for any continuous function with $f(a)f(b) < 0$.

(Recall that basic facts about convergence of sequences are summarized at the beginning of this chapter, p. 21.)

Denote the sequences of endpoints by (a_n) and (b_n) and suppose $a_0 < b_0$. Now

$$a_{n+1} = \begin{cases} a_n & \text{if } f(a_n)f(m) < 0 \\ m = \dfrac{a_n + b_n}{2} & \text{if } f(a_n)f(m) \geq 0 \end{cases}$$

It follows that (a_n) is increasing. Similarly (b_n) is decreasing.

Also $a_n < b_n \leq b_0$ for every n. That is (a_n) is increasing *and* bounded above. Therefore (a_n) converges to some limit, L, say. Similarly (b_n) is decreasing and bounded below (by a_0) and so also converges, to M, say.

Now the bisection algorithm is constructed so that

$$b_n - a_n = \frac{b_{n-1} - a_{n-1}}{2}$$

from which we may deduce that

$$b_n - a_n = \frac{b_0 - a_0}{2^n}$$

and therefore $b_n - a_n \to 0$. It follows that $L = M$; that is, the two sequences have the same limit.

Finally, because the function f is continuous,

$$f(a_n)f(b_n) \to [f(L)]^2 \geq 0$$

but the algorithm is designed to ensure that $f(a_n)f(b_n) \leq 0$ for every n. The only possibility is that $f(L) = 0$, which is to say this limit is a solution of the equation $f(x) = 0$.

Exercises: Section 2.2

1 Show that the equation

$$3x^3 - 5x^2 - 4x + 4 = 0$$

has a root in the interval $[0, 1]$. Use the bisection method to obtain an interval of length less than $1/8$ containing this solution.

How many iterations would be needed to obtain this solution with an error smaller than 10^{-6}?

2 Show that the equation $e^x - 3x - 1 = 0$ has only 1 positive solution, and that it lies in the interval $[1, 3]$. Use the bisection method to find the positive solution of this equation with error less than 10^{-5}.

3 Show that the equation

$$\exp x - 100x^2 = 0$$

has exactly 3 solutions. Obtain intervals of length less than 0.1 containing them.

2.3 Function iteration

In the previous section, we saw that bisection provides a simple and reliable technique for solving an equation which is probably sufficient for any relatively straightforward one-off problem. However, if the function f is more complicated (perhaps its values must be obtained from the numerical solution of a differential equation) or if the task is to be performed repeatedly for different values of some parameters, then a more efficient technique is needed.

If (2.1) is rewritten in the form

$$x = g(x) \tag{2.3}$$

then this *rearrangement* can be used to define an iterative process as follows. We make a guess or estimate x_0 of the solution and use this to define a sequence (x_n) by

$$x_n = g(x_{n-1}) \qquad n = 1, 2, \ldots \tag{2.4}$$

Provided that g is continuous, we see that, *if* this iterative sequence converges then the terms get closer and closer together. Eventually we obtain, to any required accuracy,

$$x_n \approx x_{n-1}$$

which is to say

$$x_{n-1} \approx g(x_{n-1})$$

so that x_n is, approximately, a solution of (2.3).

Example 4 We again consider (2.2), $x - \cos x = 0$, which we can obviously rewrite as

$$x = \cos x$$

Using the *iteration function* $g(x) = \cos x$ with the initial guess $x_0 = 0.7$ (obtained from the first few bisection iterations in Example 2, perhaps) we get

$$x_1 = \cos 0.7 = 0.764\,8$$
$$x_2 = \cos 0.764\,8 = 0.7215$$
$$x_3 = \cos 0.721\,5 = 0.750\,8$$

$$\vdots$$

$$x_9 = 0.7402$$
$$x_{10} = 0.7383$$
$$x_{11} = 0.7396$$

from which we might conclude that the solution is 0.74 to 2 decimal places.

Note that we have *proved nothing* about the convergence of this sequence.

Approaches such as this will not always work. For example, the same equation could have been rewritten as $x = \cos^{-1} x$. Using the starting value $x_0 = 0.74$, the first few iterations yield the sequence

0.7377, 0.7411, 0.7361, 0.7435, 0.7325, 0.7489, 0.7245, 0.7605,
0.7067, 0.7860, 0.6665, 0.8414, 0.5710, 0.9631, 0.2727, 1.2946,

so that after 16 iterations, the next iterate is not even defined!

To check these numbers, note that the first case is equivalent to repeatedly pressing the cos button on your calculator with the starting value 0.7 *in radians*. The second case simply uses the inverse cosine function.

These observations raise the important question: 'Can we determine, in advance, whether a particular iteration will converge?' There is a secondary aspect to this question, too. 'Can we obtain any estimates of the accuracy of our solution?' The answers to both questions are 'Yes', as we shall see.

We begin with a graphical illustration of the process of function iteration. Solving an equation of the form $x = g(x)$ is equivalent to finding the intersection of the graphs $y = x$ and $y = g(x)$. Figure 2.2 illustrates two convergent function iterations. In each case, x_n is the point at which the vertical line $x = x_{n-1}$ meets the curve $y = g(x)$. The horizontal line at this height, $y = g(x_{n-1})$ meets the graph of $y = x$ at $x = x_n = g(x_{n-1})$. The process can be continued indefinitely.

In case (a), where g is a decreasing function, the iterates converge in an oscillatory manner to the solution – giving a web-like picture. This can be very useful in determining the accuracy of the computed solution. In case (b), the iterates converge monotonically – either ascending or descending a staircase.

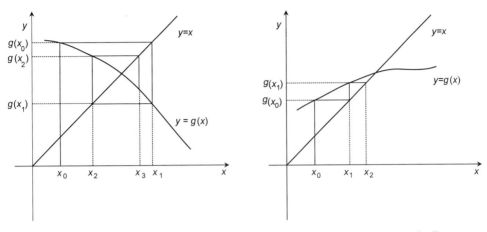

g is a decreasing function   **Iterates converge monotonically**

Figure 2.2 Convergent function iterations

In Figure 2.2, we see that function iteration can converge independent of whether the iteration function is increasing or decreasing near the solution. What turns out to be critical is *how fast* the function changes.

We shall assume that (2.3) is a rearrangement of (2.1) and that this equation has a solution s. For the purpose of our analysis, we shall denote by (e_n) the sequence of errors in the iterates. That is, for each n, we write

$$e_n = x_n - s$$

Now, using a Taylor series expansion of g about the solution, and noting that $g(s) = s$ we get

$$e_{n+1} = x_{n+1} - s = g(x_n) - g(s)$$

$$= \left[g(s) + (x_n - s)g'(s) + \frac{(x_n - s)^2}{2!}g''(s) + \ldots \right] - g(s)$$

$$= e_n g'(s) + \frac{e_n^2}{2!}g''(s) + \cdots \tag{2.5}$$

If $|e_n|$ is small enough that higher-order terms can be neglected, we get

$$e_{n+1} \approx e_n g'(s) \tag{2.6}$$

from which we deduce that the error will be reduced if $|g'(s)| < 1$. Unfortunately this is not a useful test for convergence as it stands – since we do not know s.

Theorem 1 establishes that a sufficient condition for convergence of an iteration is that $|g'(x)| < 1$ throughout an interval containing the solution.

Theorem 1   *Suppose that g is differentiable on $[a, b]$ and that*

(i) $g([a, b]) \subseteq [a, b]$ *(That is, $g(x) \in [a, b]$ for every $x \in [a, b]$), and*
(ii) $|g'(x)| \leq K < 1$ *for all $x \in [a, b]$.*

Then the equation $x = g(x)$ has a unique solution in the interval $[a, b]$ and the iterative sequence defined by

$$x_0 \in [a, b]; \quad x_n = g(x_{n-1}), \; n = 1, 2, \ldots$$

converges to this solution

Proof   Firstly, since $g(a), g(b) \in [a, b]$, it follows that

$$a - g(a) \leq 0 \leq b - g(b)$$

By the intermediate value theorem, we deduce that there exists $s \in [a, b]$ such that $s = g(s)$. The fact that s is the only such solution follows from the mean value theorem.

Suppose that $t = g(t)$ for some $t \in [a, b]$. By the mean value theorem, $g(s) - g(t) = (s - t)g'(\xi)$ for some ξ between s and t. But, $g(s) = s$ and $g(t) = t$ so that

$$s - t = (s - t)g'(\xi)$$

or

$$(s - t)[1 - g'(\xi)] = 0$$

By condition *(ii)*, $|g'(\xi)| < 1$ which implies $s - t = 0$; proving that the solution is unique.

The convergence of the iteration is also established by appealing to the mean value theorem. By condition *(i)*, we know that $x_n \in [a, b]$ for every n since $x_0 \in [a, b]$. Then, for some $\xi_n \in [a, b]$,

$$\begin{aligned}
|e_{n+1}| &= |g(x_n) - g(s)| \\
&= |x_n - s| \cdot |g'(\xi_n)| \\
&= |e_n| \cdot |g'(\xi_n)| \leq K|e_n|
\end{aligned}$$

In turn we obtain $|e_{n+1}| \leq K^{n+1}|e_0| \to 0$ which implies $x_n \to s$, as desired. ∎

Under the same conditions, if we choose $x_0 = (a + b)/2$, then $|e_0| = |x_0 - s| \leq (b - a)/2$ so that this result also provides bounds for the errors. In the next examples we see how to use this result to determine in advance whether an iteration will converge, and to estimate the accuracy of our results.

Example 5   **Consider the equation of Example 3, $e^x - 2x - 1 = 0$ which we already know has a solution in $[1, 2]$. Three possible rearrangements are**

(i) $x = \dfrac{e^x - 1}{2}$,   (ii) $x = e^x - x - 1$, and   (iii) $x = \ln(2x + 1)$

We begin by examining the first few iterations using each of these rearrangements with the iteration functions stored as m-files. In each case we used the initial guess $x_1 = 1.5$. The MATLAB commands used for the first of these were as follows; the modifications required for the other cases should be apparent:

```
» x(1)=1.5;
» for k=2:8
   x(k)=iter1(x(k-1));
  end
```

where the m-file iter1.m contains

```
function y=iter1(x)
y=(expx(x)-1)/2;
```

Note the use of expx rather than the built-in exp function. This is because there is an error in the MATLAB exponential function in versions 5.1 and 5.2 such that it frequently returns erroneous values for large arguments. (In fact it was the second of these examples which led to the discovery of this bug!) For scalar arguments, expx is defined by

```
function y=expx(x)
% Modified exponential function to avoid MATLAB bug
if abs(x) <800, y=exp(x);
elseif x<-800, y=0;
else y=Inf;
end
```

The results for the three rearrangements are:

Iteration number	Rearrangement 1	Rearrangement 2	Rearrangement 3
1	1.5	1.5	1.5
2	1.7408	1.9817	1.3863
3	2.3511	4.2733	1.3278
4	4.7484	66.485	1.2962
5	57.202	7.4798e+028	1.2788
6	3.4799e+024	Inf	1.2691
7	Inf	NaN	1.2636
8	Inf	NaN	1.2605

It is immediately apparent that the first two rearrangements are failing, while the third *appears* to be settling down, slowly.

Let us see what Theorem 1 tells us about this situation.

(i)   In the first case, we have $g(x) = \dfrac{e^x - 1}{2}$, from which we deduce that $g'(x) = e^x/2 > 1$ for every $x > \ln 2$. The theorem tells us we should not expect this iteration to converge. For the solution at $x = 0$, however, we see that $0 < g'(x) < 1$ for all $x < \ln 2$. Also, if $x < \ln 2$, then $g(x) \in (-1/2, 1/2)$. It follows that the conditions of the theorem are satisfied on this interval, and therefore that this iteration will converge to this solution. With $x_1 = 0.5$, the first few iterations yield:

> 0.5, 0.32436, 0.19157, 0.10558, 0.055676, 0.028627,
> 0.014521, 0.0073132

which we see is steadily (if slowly) approaching 0 as expected.

(ii)   This time, $g(x) = e^x - x - 1$, so that $g'(x) = e^x - 1$. Therefore again we have $g'(x) > 1$ for every $x > \ln 2$ so that the iteration should not be expected to converge to the solution in $[1, 2]$. Since $e^x > 0$ for every x, we also have $-1 < g'(x) < 1$ for all $x < \ln 2$. From the graph of $y = g(x)$ below we can also see that $g(x) \in [0, 0.5]$ for $x \in [-1, 0.5]$. The conditions of Theorem 1 are therefore satisfied for this interval and so we expect convergence to the solution $x = 0$ for any starting point in $[-1, 0.5]$.

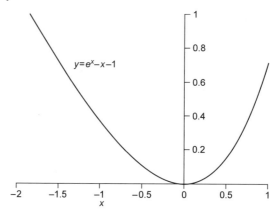

Using $x_1 = 0.5$, the first few iterates are:

> 0.5, 0.14872, 0.011628, 6.7871e−005, 2.3033e−009, 0, 0, 0

showing very rapid convergence to the solution. We shall discuss the relative speed of these two iterations shortly.

So far, we have still failed to find the desired solution in $[1, 2]$.

(iii) Using the third rearrangement, $g(x) = \ln(2x + 1)$, we obtain $g'(x) = 2/(2x + 1)$ which lies between $2/5$ and $2/3$ for $x \in [1, 2]$. It follows that g is increasing, and, since $g(1) = \ln 3$, $g(2) = \ln 5$, it follows that the conditions of the theorem are satisfied and so convergence is established for any starting point in $[1, 2]$.

The implementation of function iteration in MATLAB is easy. No additional m-files are needed beyond the iteration function itself. The greatest difficulty is knowing when to stop the iterations. In the following implementation, the iteration is stopped when either three successive iterates agree to within the tolerance (10^{-4} in this case), or some maximum number of iterations (50) has been performed.

```
» maxits=50; tol=1e-4; its=0;
» x0=1; x1=2; x2=1.5;
» while ((abs(x1-x0)>tol)|(abs(x2-x0)>tol)|(abs(x2-x1)>tol))&(its<maxits)
   x0=x1; x1=x2; x2=iter3(x1);
   its=its+1;
end
» its
its =
   15
» x2
x2 =
   1.25647620270379
```

Note: We do not need to store the full array of iterates, the last three are all that are of interest. The initial values x0=1; x1=2; are essentially arbitrary, we just need to ensure that the iterative while loop starts. The values of x0, x1 and x2 are continually overwritten as better estimates of the solution are obtained. In this case, we see that 15 iterations are sufficient to yield the accuracy prescribed. (It is common to test for convergence by examining just the last two iterates, but in the absence of other information this test is safer – even if a little too conservative.)

Note that we have *proved* nothing about the accuracy of the various iterates in Example 5. Theorem 1 can be extended to yield error estimates but these are typically pessimistic, which frequently results in performing many more iterations than are necessary. Such detailed error analysis is left to subsequent courses. We shall concentrate instead on developing methods with faster convergence rates.

In Example 5, we observed that rearrangement (ii) converged to the solution $x = 0$ much more rapidly than rearrangement (i). Why?

The answer lies in the Taylor expansion of the error in (2.5), where we see that

$$e_{n+1} = e_n g'(s) + \frac{e_n^2}{2!} g''(s) + \cdots$$

Now for iteration (i) above, we have $g'(s) = g'(0) = 1/2$ so that as the iterates converge, we expect the errors to behave so that

$$e_{n+1} \approx \frac{e_n}{2}$$

For iteration (ii), however, we have $g'(s) = g'(0) = 0$ and $g''(0) = 1$ so that the errors behave like

$$e_{n+1} \approx \frac{e_n^2}{2}$$

which will decrease much more rapidly.

This type of convergence is called *quadratic*, or *second-order convergence*. In the next section, we shall develop a general method for obtaining quadratic convergence.

Exercises: Section 2.3

1   The equation

$$3x^3 - 5x^2 - 4x + 4 = 0$$

has a solution near $x = 0.7$. (See Exercise 1, Section 2.2.) Carry out the first 5 iterations for each of the rearrangements

$$(i) \quad x = \frac{5}{3} + \frac{4}{3x} - \frac{4}{3x^2}, \qquad \text{and} \qquad (ii) \quad x = 1 + \frac{3x^3 - 5x^2}{4}$$

starting with $x_0 = 0.7$.

2   Which of the iterations in Exercise 1 will converge to the solution near 0.7? Prove your assertion using Theorem 1. Find this solution using a tolerance of 10^{-6}.

3   Find a convergent rearrangement for the solution of the equation $\exp(x) - 3x - 1 = 0$ in $[1, 3]$. Use it to locate this solution using the tolerance 10^{-6}.

4   Intervals containing the three solutions of $e^x - 100x^2 = 0$ were found in Exercise 3, Section 2.2. Each of the following rearrangements yields a convergent iteration for one of these solutions. Verify that they are all rearrangements of the original equation, and determine which will converge to which solution. Use the appropriate iterations to locate the solutions with tolerance 10^{-6}.

$$(i) \quad x = \frac{\exp(x/2)}{10}$$

$$(ii) \quad x = 2(\ln x + \ln 10)$$

$$(iii) \quad x = \frac{-\exp(x/2)}{10}$$

2.4 Newton's method

In examining the convergence of iterative schemes in Example 5, we saw that much more rapid (quadratic) convergence was achieved when $g'(s) = 0$. This is the motivation behind Newton's, or the Newton–Raphson, method. There are many descriptions of Newton's method. We shall derive it from an examination of the first-order Taylor polynomial approximation.

Suppose we wish to solve (2.1) $f(x) = 0$, as before. The first-order Taylor expansion of f about a point x_0 is

$$f(x) \approx f(x_0) + (x - x_0)f'(x_0)$$

If the point x_0 is close to the required solution s, then we expect that setting the right-hand side to zero should give a better approximation of this solution. Solving $f(x_0) + (x - x_0)f'(x_0) = 0$ yields the next approximation

$$x_1 = x_0 - \frac{f(x_0)}{f'(x_0)}$$

Repeating this argument leads to the *Newton iteration* formula

$$x_{n+1} = x_n - \frac{f(x_n)}{f'(x_n)} \qquad (n = 0, 1, 2, \ldots) \tag{2.7}$$

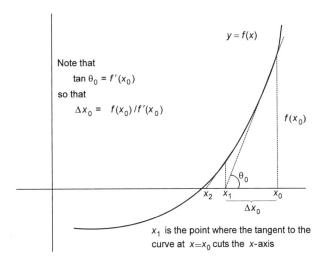

Note that
$$\tan \theta_0 = f'(x_0)$$
so that
$$\Delta x_0 = f(x_0)/f'(x_0)$$

$y = f(x)$

$f(x_0)$

θ_0

$x_2 \quad x_1 \qquad x_0$

Δx_0

x_1 is the point where the tangent to the curve at $x = x_0$ cuts the x-axis

Figure 2.3 Newton's method

Newton's method therefore uses the iteration function

$$g(x) = x - f(x)/f'(x)$$

whose derivative is given by

$$g'(x) = 1 - f'(x)/f'(x) + \frac{f(x)f''(x)}{[f'(x)]^2}$$

$$= \frac{f(x)f''(x)}{[f'(x)]^2}$$

Since $f(s) = 0$, it follows that $g'(s) = 0$ as required.

Newton's method is therefore expected to exhibit second-order convergence – *when it converges*.

There is a simple graphical representation of Newton's method (see Figure 2.3).

Newton's method is widely used in computers as a basis for square root and reciprocal evaluation.

Example 6   Find the positive square root of a real number c by Newton's method.

To find $\sqrt{c}$, we wish to solve the equation

$$x^2 - c = 0$$

for which the Newton iteration is

$$x_{n+1} = x_n - \frac{x_n^2 - c}{2x_n}$$

$$= \frac{x_n + c/x_n}{2}$$

We shall consider the general convergence properties of Newton's method shortly, but for now we simply comment that this iteration is convergent for every choice $x_0 > 0$.

For example, let $c = 5$ and choose $x_0 = 2$. Then we get

$$x_1 = \frac{2 + 5/2}{2} = 2.25$$

$$x_2 = \frac{2.25 + 5/2.25}{2} = 2.236\,111$$

$$x_3 = 2.236068$$

$$x_4 = 2.236068$$

so that after just three iterations we have agreement to 3 decimal places, and to 6 decimals after a fourth iteration.

Next we consider the implementation of Newton's method for solving $f(x) = 0$ in MATLAB. The necessary inputs are the function f, its derivative, an initial guess and the required accuracy in the solution. In the implementation below, no limit is placed on the number of iterations. This would be needed for a robust implementation of Newton's (or any iterative) method. Our objective is not to create such software, but rather to get an idea of the basic ideas. MATLAB has some robust equation-solvers built in. We shall discuss those briefly later.

| **Program** | **MATLAB m-file for Newton's method** |

```
function sol=newton(fcn,df,g,tol)
% Newton's method for solution of the equation
% fcn(x)=0
% to an accuracy tol<1. The initial guess is input as g.
% fcn, and its derivative df must be saved in m-files
old=g+1; % Chosen to ensure that the iteration starts.
while abs(g-old)>tol
    old=g;
    g=old-feval(fcn,old)/feval(df,old);
end
sol=g;
```

Example 7 Solve the equation $e^x - 2x - 1 = 0$ using Newton's method

We know the positive solution lies in $[1, 2]$ and so take the initial guess 1.5. The function $e^x - 2x - 1$ and its derivative were saved in m-files called eq1.m and deq1.m, respectively after which the following MATLAB command gave the result shown.

```
» s=newton('eq1','deq1',1.5,1e-10)
s =
    1.25643120862617
```

To get an idea of the speed with which Newton's method found this solution, the ; was removed from the line generating the new value of g so that each iteration is printed in the MATLAB command window. The successive iterates were:

```
g=1.30590273129081, 1.25905872840914, 1.25643918929784,
    1.25643120870011, and 1.25643120862617
```

Just five iterations were needed to obtain the result to high precision.

Newton's method will not always work, however. Consideration of (2.7) shows that if $f'(x_n)$ is very small, then the correction made to the iterate x_n will be very large. Thus, if the derivative of f is zero (or very small) near the solution, then Newton's method is unlikely to be successful. Such a situation can arise if there are two solutions very close together – or, as in Figure 2.4, when the gradient of f is small everywhere except very close to the solution.

Example 8   Consider the equation $\tan^{-1}(x-1) = 0.5$ which has its unique solution $x = 1.5463$ to four decimals

The function $f(x) = \tan^{-1}(x-1) - 0.5$ is graphed in Figure 2.4, along with the first 2 iterations of Newton's method with the rather poor starting point $x_0 = 4$ which yields $x_1 \approx -3.5$ and $x_2 \approx 36$. The oscillations get steadily wilder reaching Inf and then NaN in just eight more iterations using MATLAB.

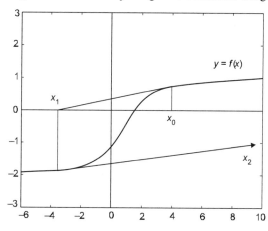

Figure 2.4   Newton's method can fail

Of course, this particular equation can be solved very easily: rewrite it in the form $x = 1 + \tan(1/2) = 1.546\,302\,5$ to obtain the 'solution'. This rearrangement transforms the original problem from one of equation-solving to one of function evaluation, which is still a real computational problem, as we shall see in later chapters.

In Example 8, we see that Newton's method can fail. The natural question is now 'Can we determine when Newton's method will converge?' The answer is 'Yes'. The situation will be summarized in the global convergence theorem for Newton's method which is stated below but not proved here.

In Figure 2.4, we see that the function f is convex (concave up) to the left of the solution and concave (concave down) to the right. Obviously therefore it has a point of inflection close to the solution. This is the source of the difficulty. The global convergence theorem, Theorem 2, has hypotheses which eliminate the possibility of any points of inflection in the interval.

Theorem 2 *Let f be twice differentiable on an interval $[a,b]$ and satisfy the conditions*

(i) $f(a)f(b) < 0$
(ii) f' *has no zeros on* $[a,b]$
(iii) f'' *does not change sign in* $[a,b]$, *and*
(iv) $\left|\dfrac{f(a)}{f'(a)}\right|, \left|\dfrac{f(b)}{f'(b)}\right| < b - a$

Then $f(x) = 0$ has a unique solution $s \in (a,b)$ and Newton's iteration will converge to s from any starting point in $[a,b]$.

Proof The first condition establishes the existence of a solution since f must change sign in the interval. The second and third conditions force both f and f' to be strictly monotone. These guarantee the uniqueness of the solution and the absence of any inflection points. The final condition ensures that a Newton iteration from either endpoint a or b will generate a point in the interval (a,b). In combination these last two conditions ensure that all the iterates remain in the interval. The details of the proof are omitted. ■

Our motivation for Newton's method was the desire to achieve quadratic convergence. In Example 9, we verify this from the error analysis of Newton's method for square roots, as in Example 6.

Example 9 We show that the iteration $x_{n+1} = \dfrac{x_n + c/x_n}{2}$ converges quadratically to $\sqrt{c}$ for any $x_0 > 0$

We consider the sequence of errors defined by

$$e_n = x_n - \sqrt{c} \quad (n = 0, 1, 2, \ldots)$$

Now $x_{n+1} = (x_n^2 + c)/2x_n$ and so

$$x_{n+1} - \sqrt{c} = \frac{x_n^2 - 2x_n\sqrt{c} + c}{2x_n} = \frac{(x_n - \sqrt{c})^2}{2x_n}$$

Similarly

$$x_{n+1} + \sqrt{c} = \frac{(x_n + \sqrt{c})^2}{2x_n}$$

from which we may deduce that

$$\frac{x_{n+1} - \sqrt{c}}{x_{n+1} + \sqrt{c}} = \frac{(x_n - \sqrt{c})^2}{(x_n + \sqrt{c})^2} = \left(\frac{x_{n-1} - \sqrt{c}}{x_{n-1} + \sqrt{c}}\right)^4$$

$$= \cdots = \left(\frac{x_0 - \sqrt{c}}{x_0 + \sqrt{c}}\right)^{2^{n+1}}$$

Since $x_0 > 0$, it follows that $\left|\dfrac{x_0 - \sqrt{c}}{x_0 + \sqrt{c}}\right| < 1$ and hence $\dfrac{x_{n+1} - \sqrt{c}}{x_{n+1} + \sqrt{c}} \to 0$ as $n \to \infty$. Therefore $x_n \to \sqrt{c}$ for every choice $x_0 > 0$.

Since even powers of real numbers are positive, it also follows from the analysis above that $x_n > \sqrt{c}$ for every $n \geq 1$ so that

$$e_{n+1} = \frac{e_n^2}{2x_n} \leq \frac{e_n^2}{2\sqrt{c}}$$

which is to say the convergence is quadratic.

Let $x = f \times 2^E$ be a normalized binary floating-point number, $f \in [1, 2)$. Then we can rewrite x as

$$x = \begin{cases} f \times 2^E & \text{if } E \text{ is even} \\ (2f)2^{E-1} & \text{if } E \text{ is odd} \end{cases}.$$

so that

$$\sqrt{x} = \begin{cases} \sqrt{f} \times 2^{E/2} & \text{if } E \text{ is even} \\ \sqrt{2f}\,2^{(E-1)/2} & \text{if } E \text{ is odd} \end{cases}$$

It follows that we need only find $\sqrt{c}$ for $c \in [1, 4)$. Then $\sqrt{c} \in [1, 2)$ and taking $x_0 = 3/2$ ensures that $e_0 \leq 1/2$. The analysis in Example 9 can then be used to establish that 5 iterations are sufficient to obtain full double precision floating-point accuracy in $\sqrt{c}$. This is why Newton's method is the basis of many computer square-root routines. (Of course, the implementation is more sophisticated than shown here.)

Similarly, Newton's method is often used for obtaining the reciprocal of $c > 0$ by solving

$$\frac{1}{x} - c = 0$$

The Newton iteration for this equation does not entail any division and so can be used as a basis for division of floating-point numbers. This is explored further in the exercises.

Exercises: Section 2.4

1   The equation

$$3x^3 - 5x^2 - 4x + 4 = 0$$

has a solution near $x = 0.7$. (See Exercise 1, Section 2.2.) Carry out the first 4 iterations of Newton's method to obtain this solution.

2   Use Newton's method to obtain the solution of the equation $\exp(x) - 3x - 1 = 0$ in $[1, 3]$ using the tolerance 10^{-12}.

3   Intervals containing the 3 solutions of $e^x - 100x^2 = 0$ were found in Exercise 3, Section 2.2. Use Newton's method to locate the solutions with tolerance 10^{-10}.

4   For the equation in Exercise 3, 2 solutions are close to $x = 0$. Try to find the critical value c such that if $x_0 > c$ then the solution near 0.1 is obtained, while if $x_0 < c$ the negative solution is located. Now try to justify your answers theoretically. (We are trying to find the *regions of attraction* for each of these solutions.)

5   Show that Newton's method for finding reciprocals by solving $1/x - c = 0$ results in the iteration

$$x_{n+1} = x_n(2 - cx_n)$$

Show that this iteration function satisfies $|g'(x)| < 1$ for $x \in (1/2c, 3/2c)$.

6   For the iteration in Exercise 5, prove that

$$x_{n+1} - \frac{1}{c} = -c\left(x_n - \frac{1}{c}\right)^2$$

Therefore $x_{n+1} < 1/c$. Show also that if $x_n < 1/c$ then $x_{n+1} > x_n$. It follows that, for $n \geq 1$, (x_n) is an increasing sequence which converges quadratically to $1/c$.

7   For Newton's iteration for finding $1/c$ with $c \in [1, 2)$ and $x_0 = 3/4$, show that 6 iterations will yield an error smaller than 2^{-65}.

2.5   The secant method

In the previous section, we saw that Newton's method can be a very powerful tool for solving equations. There is, however, one important potential difficulty. Implementation of Newton's method requires knowledge of the derivative of the function involved. In many practical situations this derivative may not be available – if, for example, the function itself is the result of some other computation. One important example of this is in shooting methods for the solution of differential equations. We shall consider this technique in Chapter 6.

The *secant method* is an attempt to recover some of the power of Newton's method without requiring knowledge of the derivative. We begin with a graphical description of this approach. Newton's method was described by the idea that the next iterate should be the point at which the tangent line (at the current estimate of

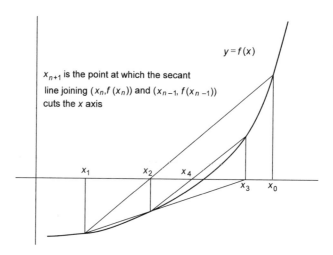

Figure 2.5   The secant method

the solution) cuts the x-axis. The analogous description for the secant method is that the next iterate is the point at which the secant, or chord, line joining the previous 2 iterates cuts this axis. This is illustrated in Figure 2.5.

To obtain an algebraic description of this iteration, we begin with the equation of the secant line joining 2 points on the curve $y = f(x)$ at $x = x_0, x_1$. This line has slope $\dfrac{f(x_1) - f(x_0)}{x_1 - x_0}$ and hence its equation can be written in the form

$$y - f(x_1) = \frac{f(x_1) - f(x_0)}{x_1 - x_0}(x - x_1) \tag{2.8}$$

Setting $y = 0$ in (2.8) and denoting the solution by x_2 we get

$$x_2 = x_1 - \frac{x_1 - x_0}{f(x_1) - f(x_0)}f(x_1)$$

The general iteration formula for the secant method is therefore

$$x_{n+1} = x_n - \frac{x_n - x_{n-1}}{f(x_n) - f(x_{n-1})}f(x_n) \tag{2.9}$$

Before considering any examples, we observe that this formula is just the Newton iteration with the true derivative replaced by the approximation

$$f'(x_n) \approx \frac{f(x_n) - f(x_{n-1})}{x_n - x_{n-1}}$$

As a general rule, we shall see that approximate numerical differentiation is unreliable. In this context, however, this simple approximation is adequate. A true value of the slope is not needed here and this approximation provides enough information to make the secant method highly efficient in many cases.

Example 10 **Apply the secant method to the solution of**
$$e^x - 2x - 1 = 0$$
in the interval $[1, 2]$

The function $e^x - 2x - 1$ has already been used in the m-file eq1.m. The first few iterations are easily performed by the following sequence of MATLAB commands:

```
» x0=1; x1=2;
» x2=x1-(x1-x0)/(eq1(x1)-eq1(x0))*eq1(x1)
x2 =
   1.10548183523245
» x0=x1; x1=x2;
» x2=x1-(x1-x0)/(eq1(x1)-eq1(x0))*eq1(x1)
x2 =
   1.17147257557915
```

The last two commands can be repeated to generate the subsequent iterates:

$$1.27498830910932, \ 1.25449558049622, \ 1.25639007263928$$

which can be seen to be settling down quite rapidly. After just four more iterations, we have two successive iterates which are both 1.25643120862617 to the full precision of MATLAB.

Like Newton's method, the secant method appears to converge very rapidly once we are close to the solution. There are convergence theorems for this method similar to those for Newton's method, but they are beyond the scope of this book. The main conclusion is that when the secant method converges, it does so at a *superlinear* rate. Specifically, if we again use e_n to denote the error $x_n - s$, then the errors eventually satisfy

$$e_{n+1} \approx c e_n^\alpha$$

where $\alpha = \left(1 + \sqrt{5}\right)/2 \approx 1.6$. This can be interpreted as saying that the number of correct decimal places increases by about 60% with each iteration. (The corresponding statement for Newton's method is that this number of correct decimal places doubles every iteration.)

Example 11 **The secant method can be applied to computing square roots**

If $f(x) = x^2 - c$ then the secant iteration becomes

$$x_{n+1} = x_n - \left(x_n^2 - c\right) \frac{x_n - x_{n-1}}{x_n^2 - x_{n-1}^2} = x_n - \frac{x_n^2 - c}{x_n + x_{n-1}}$$

$$= \frac{x_n^2 + x_n x_{n-1} - x_n^2 + c}{x_n + x_{n-1}} = \frac{x_n x_{n-1} + c}{x_n + x_{n-1}}$$

Recall that Newton's iteration for this equation is

$$x_{n+1} = \frac{x_n^2 + c}{2x_n} = \frac{x_n x_n + c}{x_n + x_n}$$

We see great similarity between these formulas – especially when the iterates are close to the solution.

For $c = 5$, with the initial guesses $x_0 = 2$, $x_1 = 5/2$, the next few secant iterations are: 2.22222222222222, 2.23529411764706, 2.23607038123167, 2.23606797708378, 2.23606797749979 and 2.23606797749979. 6 iterations are sufficient to give full MATLAB accuracy.

For Newton's method, the corresponding iterates starting at $x_0 = 2$ are: 2.25, 2.23611111111111, 2.2360679779158, 2.23606797749979 and 2.23606797749979 so that just 5 iterations are sufficient.

The price paid for using the secant method – and no derivative information – appears to be very slight in this case.

Writing MATLAB code for the secant method is left as an exercise. Use the Newton's method program as a guide.

Exercises: Section 2.5

1 The equation

$$3x^3 - 5x^2 - 4x + 4 = 0$$

has a solution near $x = 0.7$. (See Exercise 1, Section 2.2.) Carry out the first 4 iterations of the secant method to obtain this solution. Compare the results with those of Newton's method in Exercise 1, Section 2.4.

2 Use the secant method to obtain the solution of the equation $\exp(x) - 3x - 1 = 0$ in $[1, 3]$ using the tolerance 10^{-12}.

3 Write a MATLAB program (m-file) to solve an equation using the secant method. Test it by checking your answers to the previous exercises.

4 Intervals containing the 3 solutions of $e^x - 100x^2 = 0$ were found in Exercise 3, Section 2.2. Use the secant method to locate the solutions with tolerance 10^{-10}. Compare the numbers of iterations used with those needed by Newton's method.

5 Show that the secant method for finding reciprocals by solving $1/x - c = 0$ results in a division-free iteration

$$x_{n+1} = x_n + x_{n-1} - cx_n x_{n-1}$$

Carry out the first 4 iterations for finding $1/7$ using $x_0 = 0.1$, $x_1 = 0.2$.

2.6 2 equations in 2 unknowns: Newton's method

In this section we consider the problem of solving a system of equations

$$\mathbf{f}(\mathbf{x}) = \mathbf{0} \tag{2.10}$$

where $\mathbf{f}$ is a vector function of the vector variable $\mathbf{x}$. Mostly we shall specialize the situation to just 2 equations in 2 unknowns, but we begin with the more general case of n equations in n unknowns.

Newton's method for such a system is again based on a first-order Taylor expansion:

$$\mathbf{f}(\mathbf{x} + \mathbf{h}) \approx \mathbf{f}(\mathbf{x}) + J\mathbf{h} \tag{2.11}$$

where J is the Jacobian matrix of $\mathbf{f}$ at $\mathbf{x}$:

$$J_{ij} = \frac{\partial f_i}{\partial x_j}$$

evaluated at $\mathbf{x}$. Here f_i represents the i-th equation, or component of $\mathbf{f}$. As usual, x_j is the j-th component of the vector $\mathbf{x}$.

As in the 1-dimensional case, the Newton iteration is derived from setting the right-hand side of (2.1) to $\mathbf{0}$. This leads to the iterative formula

$$\mathbf{x}_{k+1} = \mathbf{x}_k - J^{-1}\mathbf{f}(\mathbf{x}_k) \tag{2.12}$$

though this is not usually the way it is implemented.

Implementation of the iteration (2.12) requires the ability to solve general linear systems of equations. This is the first time (of many) that we shall see the need for this ability. The solution of linear systems of equations is discussed in Chapter 7. The methods discussed there can be used to implement Newton's method for a general system. For the remainder of this section, we concentrate on the case $n = 2$ – that is 2 equations in 2 unknowns.

We wish to solve the system

$$f_1(x,y) = 0 \tag{2.13}$$
$$f_2(x,y) = 0$$

The first-order Taylor expansions of these functions gives

$$f_1(x+h, y+k) \approx f_1(x,y) + hf_{1x}(x,y) + kf_{1y}(x,y)$$
$$f_2(x+h, y+k) \approx f_2(x,y) + hf_{2x}(x,y) + kf_{2y}(x,y)$$

Setting the two right-hand sides (simultaneously) to zero leads to the solutions

$$
\begin{aligned}
h &= \frac{-f_1 f_{2y} + f_2 f_{1y}}{f_{1x} f_{2y} - f_{1y} f_{2x}} \\
k &= \frac{f_1 f_{2x} - f_2 f_{1x}}{f_{1x} f_{2y} - f_{1y} f_{2x}}
\end{aligned}
\tag{2.14}
$$

where all functions are to be evaluated at (x,y).

The Newton iteration for a pair of equations in 2 unknowns is therefore

$$x_{n+1} = x_n + h; \quad y_{n+1} = y_n + k \tag{2.15}$$

where (h, k) are given by (2.14) evaluated at (x_n, y_n).

Example 12 **Find the coordinates of the intersections in the first quadrant of the ellipse $4x^2 + y^2 = 4$ and the curve $x^2 y^3 = 1$ illustrated below**

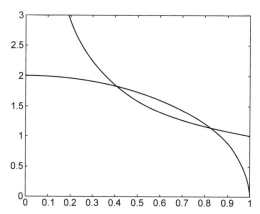

One possible technique for solving this pair of equations would be to eliminate x^2 between the two equations to obtain a fifth-degree polynomial in y. This could in turn be solved using Newton's method for a single equation. Our purpose here is to illustrate the use of Newton's method for a pair of equations. The use of this substitution is left as an exercise to serve as a check on this process.

One of the solutions is close to $(0.4, 1.8)$ and we shall use this as a starting point for the iteration. The partial derivatives of the 2 functions $f_1(x, y) = 4x^2 + y^2 - 4$ and $f_2(x, y) = x^2 y^3 - 1$ are

$$f_{1x} = 8x, \quad f_{1y} = 2y$$
$$f_{2x} = 2xy^3, \quad f_{2y} = 3x^2 y^2$$

At the point $\mathbf{x}_0 = (0.4, 1.8)$, we have

$$f_1 = 4(0.4)^2 + 1.8^2 - 4 = -0.12$$
$$f_2 = (0.4)^2 (1.8)^3 - 1 = -0.066\,88$$

and

$$f_{1x} = 3.2, \quad f_{1y} = 3.6$$
$$f_{2x} = 4.665\,6, \quad f_{2y} = 1.555\,2$$

Hence $f_{1x}f_{2y} - f_{1y}f_{2x} = (3.2)(1.5552) - (3.6)(4.6656) = -11.81952.$ Now applying (2.15) we get

$$h = \frac{-(-0.12)(1.5552) + (-0.06688)(3.6)}{-11.81952} = 4.5808967 \times 10^{-3}$$

$$k = \frac{(-0.12)(4.6656) - (-0.06688)(3.2)}{-11.81952} = 2.9261425 \times 10^{-2}$$

so that

$$x_1 = 0.4 + 4.5808967 \times 10^{-3} = 0.4045809$$
$$y_1 = 1.8 + 2.9261425 \times 10^{-2} = 1.8292614$$

Subsequent iterations generate the points

$(0.404149564420627, 1.82938603854765),$
$(0.404149457020688, 1.82938592581215),$
$(0.404149457020644, 1.82938592581218).$

A typical iteration of this method for this pair of equations can be implemented in the MATLAB command window using:

```
» f1=4*x0^2+y0^2-4; f2=x0^2*y0^3-1;
» f1x=8*x0; f1y=2*y0; f2x=2*x0*y0^3; f2y=3*x0^2*y0^2;
» D=f1x*f2y-f1y*f2x;
» h=(f2*f1y-f1*f2y)/D; k=(f1*f2x-f2*f1x)/D;
» x0=x0+h;y0=y0+k;
```

The other solution is close to $(0.8, 1.2)$. Using this starting point, we get the iterates

$(0.822986111111111, 1.1387037037037),$
$(0.821684259596287, 1.13988951023044),$
$(0.821681625203841, 1.13989351575193),$
$(0.821681625190098, 1.13989351577235),$ and
$(0.821681625190098, 1.13989351577235)$ again.

A very small number of iterations has provided both solutions to very high accuracy.

Newton's method is fairly easy to implement for the case of two equations in two unknowns. We first need the function m-files for the equations and the partial derivatives. For the equations in Example 12, these can be given by:

```
function f=eq2(v)
%Here both f and v are vector quantities
x=v(1);y=v(2);
f(1)=4*x^2+y^2-4;
f(2)=x^2*y^3-1;
```

and

```
function J=Deq2(v)
% Jacobian matrix for eq2.m
x=v(1);y=v(2);
J(1,1)=8*x; J(1,2)=2*y;
J(2,1)=2*x*y^3; J(2,2)=3*x^2*y^2;
```

The Newton2.m m-file will need both the function and its partial derivatives as well as a starting vector and a tolerance. The following code can be used.

Program

MATLAB m-file for Newton's method for 2 equations

```
function sol=newton2(fcn,Jac,g,tol)
% Newton's method for solution of two equations given by
% fcn(x)=0, where fcn is a vector function
% to an accuracy tol% <1. The initial guess is input as g.
% fcn, and its partial derivatives Jac must be saved in m-files
old(1)=g(1)+1;
while max(abs(g-old))>tol
   old=g;
   f=feval(fcn,old); f1=f(1); f2=f(2);
   J=feval(Jac,old);
   f1x=J(1,1); f1y=J(1,2);
   f2x=J(2,1); f2y=J(2,2);
   D=f1x*f2y-f1y*f2x;
   h=(f2*f1y-f1*f2y)/D; k=(f1*f2x-f2*f1x)/D;
   g=old+[h,k];
end
sol=g;
```

Then the following command can be used to generate the second of the solutions in Example 2:

```
» s=Newton2('eq2','deq2',[0.8,1.2],1e-8)
s =
   0.821681625190098 1.13989351577235
```

Exercises: Section 2.6

1 Eliminate x^2 between the equations $4x^2 + y^2 = 4$ and $x^2y^3 = 1$ to get a polynomial equation in y for the intersection points of these curves (see Example 12.)

2 Use appropriate starting points to find the y coordinates of the intersection points of the curves in Exercise 1 using Newton's method for a single equation. Verify that these solutions yield the same intersection points as were found in Example 12.

3   In the figure below are graphs of the curves defined by the equations
$x^4 + xy^2 + y^4 = 1$ and $x^2 + xy - y^2/4 = 1$

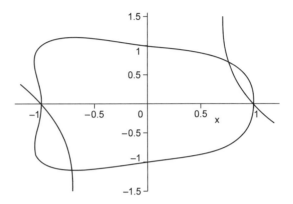

There are 2 obvious intersection points at $(\pm 1, 0)$. Use an appropriate starting
point and perform the first 2 iterations of Newton's method for the other
intersection in the first quadrant.

4   For the curves in Exercise 3, there are intersection points close to $(0.8, 0.8)$ and
$(-0.75, -1)$. Use Newton's method to find these intersection points to high
accuracy.

2.7   MATLAB functions for equation solving

For many of the tasks we shall discuss including the solution of equations,
MATLAB has built-in functions using efficient algorithms.

fzero   In the case of solving nonlinear equations of a single variable, the basic MATLAB
function is fzero.

This m-file takes two required arguments, the function f and the starting point x_0.
It then finds the solution of $f(x) = 0$ nearest to x_0. Optional arguments are the
tolerance, and 'trace'. If the latter is nonzero then MATLAB shows the result of
each iteration.

The method used is a combination of the bisection method (Section 2.2), the
secant method (Section 2.5), and a more advanced technique, *inverse quadratic
interpolation*. We shall discuss polynomial interpolation in Chapter 4, though this
particular technique is beyond the scope of this book. (Newton's method cannot be
used in this general context because of its need for the derivative.)

Example 13   We use MATLAB's fzero function to solve $e^x - 2x - 1 = 0$ with tolerance 10^{-8} (recall that this function has already been used. It was saved as eq1.m)

The command

> » s=fzero('eq1',1,1e-8)

gives the output

> s =
>   1.2564312086028

For more details on this and other MATLAB functions, consult your MATLAB manual or type help fzero at the MATLAB prompt.

---

roots   For the special case of finding roots of polynomial equations, we can use roots.

This m-file will find *all* roots of the polynomial equation $p(x) = 0$ where p is represented in MATLAB by the vector of its coefficients, beginning with the highest degree term. The polynomial $x^2 + 2x - 3$ is therefore represented by the vector $[1, 2, -3]$. Any zero coefficients must be included.

Example 14   Solve the polynomial equation $y^5 - 4y^3 + 4 = 0$ (see Exercise 1, Section 2.6)

The polynomial is represented by the coefficient vector $[1, 0, -4, 0, 0, 4]$. The MATLAB command

> » roots([1,0,-4,0,0,4])

gives the output

> ans =
>   1.82938592581218
>   -2.1045484953753
>   1.13989351577235
>   -0.432365473104611 + 0.85117984789112i
>   -0.432365473104611 - 0.85117984789112i

Notice that the real and complex roots are all listed. Note too that the two positive real roots coincide with our Newton's method solution to the original curve intersection problem in Example 12.

For systems of nonlinear equations there is no simple equivalent of fzero. However, there is the function fmins which implements an algorithm for the minimization of a function of several variables. The connection between these two problems is that solving a system of equations $f_1(\mathbf{x}) = f_2(\mathbf{x}) = \cdots = f_n(\mathbf{x}) = 0$ is equivalent to locating the minimum of the function

$$F(\mathbf{x}) = \sum_{k=1}^{n} |f_k(\mathbf{x})|^2$$

The details of the algorithms used are topics for more advanced textbooks.

3 Approximate Evaluation of Functions

Aims and objectives

In this chapter, we address the question of how to obtain good *approximations* to some of the standard elementary functions. We introduce both the traditional approach using series expansions and its modern equivalent, the CORDIC algorithms used by almost all calculators and personal computer chips. With easy access to so many functions at the touch of a button, one of the primary objectives is to convince the reader that there is a problem to be solved – as well as giving an introduction to two powerful techniques.

3.1 Introduction

The next several chapters are concerned with computing functional quantities beginning with the evaluation of some of the basic functions of mathematics. The elementary functions such as sine, cosine, exponential and logarithm functions are not computable *exactly* except in very special cases. Some are even defined in terms of integrals which cannot be computed exactly by standard algebra or calculus techniques. For example, the natural logarithm function is often defined as

$$\ln x = \int_1^x \frac{1}{t} dt$$

Techniques for numerical evaluation of integrals such as this will be discussed in Chapter 5.

Other elementary functions, such as arctan, are defined as inverses of functions which themselves cannot be readily evaluated exactly. Implicitly, this defines them as solutions of nonlinear equations which cannot be solved algebraically. Techniques such as those used in Chapter 2 are then potential methods of evaluating such functions.

In many cases there are better techniques available. Many of these are based on either the use of series expansions (Section 3.2) or the CORDIC algorithms (Section 3.3) which are widely used in calculators and personal computer (PC) hardware. If you consult the MATLAB manual to see what methods are used for some of the elementary and special functions of mathematics, you will see other techniques mentioned, such as rational approximation and asymptotic series. The specifics of these methods are beyond our scope here but many of the principles used here are also employed in those techniques to ensure the desired accuracy in the resulting approximations.

In the case of series expansions, the important theoretical issues are finding the radius of convergence of the series expansion, and then determining the number of terms that are needed in order to reduce the truncation error below some prescribed tolerance. Once that is achieved, the summation of the appropriate terms of the series is usually a straightforward task.

For both series expansions and the CORDIC algorithms there is further difficulty in handling arguments outside their intervals of convergence. These problems can often be overcome by some *range reduction* method. As an (apparently) simple example of this, we know that the basic trigonometric functions sin and cos are periodic with period 2π. It is sufficient therefore to have good algorithms for computing these functions for arguments in the interval $[0, 2\pi]$; larger arguments can be reduced to this interval by subtracting an appropriate multiple of 2π. As we shall see there is much more to it than this implies.

3.2 Series expansions

There are two absolutely fundamental power series from which many of the other important examples are derived: the *geometric series*

$$\frac{1}{1-x} = \sum_{k=0}^{\infty} x^k = 1 + x + x^2 + \cdots \qquad (|x| < 1) \qquad (3.1)$$

and the *exponential series*

$$\exp(x) = e^x = \sum_{k=0}^{\infty} \frac{x^k}{k!} = 1 + x + \frac{x^2}{2!} + \frac{x^3}{3!} + \cdots \qquad (\text{all } x) \qquad (3.2)$$

Other important series expansions are easily derived from these, or by Taylor or MacLaurin expansions. Using the identity

$$e^{ix} = \cos x + i \sin x$$

where i is the imaginary $\sqrt{-1}$, we get the series for the two basic trigonometric functions:

$$\cos x = \sum_{k=0}^{\infty} \frac{(-1)^k x^{2k}}{(2k)!} = 1 - \frac{x^2}{2!} + \frac{x^4}{4!} \cdots \qquad (\text{all } x) \qquad (3.3)$$

$$\sin x = \sum_{k=0}^{\infty} \frac{(-1)^k x^{2k+1}}{(2k+1)!} = x - \frac{x^3}{3!} + \frac{x^5}{5!} \cdots \qquad (\text{all } x) \qquad (3.4)$$

(Alternatively, if you are unfamiliar with complex numbers, these are the MacLaurin series for these functions.) Series expansions for the hyperbolic functions can be obtained in a similar manner

$$\cosh x = \frac{1}{2}(e^x + e^{-x}) = \sum_{k=0}^{\infty} \frac{x^{2k}}{(2k)!} = 1 + \frac{x^2}{2!} + \frac{x^4}{4!} + \cdots \qquad \text{(all } x) \qquad (3.5)$$

$$\sinh x = \frac{1}{2}(e^x - e^{-x}) = \sum_{k=0}^{\infty} \frac{x^{2k+1}}{(2k+1)!} = x + \frac{x^3}{3!} + \frac{x^5}{5!} + \cdots \qquad \text{(all } x) \qquad (3.6)$$

By integrating the power series (3.2) we get

$$\ln(1-x) = -\sum_{k=0}^{\infty} \frac{x^{k+1}}{k+1} = -x - \frac{x^2}{2} - \frac{x^3}{3} - \cdots \qquad (|x| < 1) \qquad (3.7)$$

and, replacing x by $-x$,

$$\ln(1+x) = -\sum_{k=0}^{\infty} \frac{(-1)^{k+1} x^{k+1}}{k+1} = x - \frac{x^2}{2} + \frac{x^3}{3} - \frac{x^4}{4} \cdots \qquad (|x| < 1) \qquad (3.8)$$

For more details on the derivations of these formulas, consult your calculus text.

The rest of this section is devoted to some important and illustrative examples.

Example 1   **The series in (3.8) is convergent for $x = 1$. It follows that**

$$\ln 2 = 1 - \frac{1}{2} + \frac{1}{3} - \frac{1}{4} \cdots$$

Use the first 8 terms of this series to estimate ln 2. How many terms would be needed to compute ln 2 with an error less than 10^{-6} using this series? (Note: The true value of $\ln 2 \approx 0.693\,147\,18$)

The first 8 terms yield

$$\ln 2 \approx 1 - \frac{1}{2} + \frac{1}{3} - \frac{1}{4} + \frac{1}{5} - \frac{1}{6} + \frac{1}{7} - \frac{1}{8} = 0.634\,523\,81$$

which has an error close to 0.06.

Since the series (3.8) is an alternating series of decreasing terms (for $0 < x \le 1$), the truncation error is smaller than the first term omitted. To force this truncation error to be smaller than 10^{-6} would therefore require that the first term omitted is smaller than $1/1\,000\,000$. That is, the first one million terms would suffice.

This is obviously not a practical approach. We shall return to the natural logarithm function later.

Example 2 **Derive a series expansion for the function arctan x. Use this series and the identity arctan $1 = \pi/4$ to compute π.**

First, we know that

$$\frac{d}{dt}\arctan t = \frac{1}{1 + t^2}$$

and so, for $|t| < 1$, using (3.1) with $x = -t^2$ we get

$$\frac{d}{dt}\arctan t = 1 - t^2 + t^4 - t^6 \cdots$$

Power series may be integrated term-by-term within their radius of convergence. Hence, for $|x| < 1$,

$$\begin{aligned}
\arctan x &= \int_0^x \left(1 - t^2 + t^4 - t^6 \cdots\right) dt \\
&= x - \frac{x^3}{3} + \frac{x^5}{5} - \frac{x^7}{7} \cdots
\end{aligned} \tag{3.9}$$

This series is also convergent for $x = 1$ and so we may deduce that

$$\frac{\pi}{4} = \arctan 1 = 1 - \frac{1}{3} + \frac{1}{5} - \frac{1}{7} \cdots$$

The first eight terms yield the approximation

$$\pi \approx 4 \sum_{k=0}^{7} \frac{(-1)^k}{2k + 1} = 3.017\,071\,8$$

Adding the next term we get $\pi \approx 3.017\,071\,8 + 4/17 = 3.252\,365\,9$. It is apparent that many more terms will be needed to obtain a good approximation to π using this series.

For *single*-precision floating-point, we would require an error in the series approximation of $\pi/4$ smaller than 2^{-25}. Since the series is alternating with decreasing terms, we could stop once the first term omitted is smaller than this bound: $2^{24} = 16\,777\,216$ would suffice! For IEEE double precision, as is used by MATLAB, the required number of terms rises to $2^{53} \approx 10^{16}$. Even on a very fast *gigaflop* (10^9 floating-point operations per second) computer, we would be waiting about 10 million seconds, or nearly 4 months to obtain a value for π. Typing pi at the MATLAB prompt should convince you that better techniques must be available! We shall consider some improvements shortly.

These first two examples using series approximations make it plain that finding an elegant mathematical expression for a quantity is not the same as finding a good algorithm for its evaluation!

Example 3   Find the number of terms of the exponential series that are needed for exp x to have error $< 10^{-5}$ for $|x| \leq 2$

First we observe that the tail of the exponential series truncated after N terms increases with $|x|$. Also the truncation error for $x > 0$ will be greater than that for $-x$ since the series for $\exp(-x)$ will be alternating in sign. It is sufficient therefore to consider $x = 2$.

We shall denote by $E_N(x)$ the truncation error in the approximation using N terms:

$$\exp x \approx \sum_{k=0}^{N-1} \frac{x^k}{k!}$$

Then, we obtain, for $x > 0$

$$\begin{aligned}
E_N(x) &= \sum_{k=N}^{\infty} \frac{x^k}{k!} = \frac{x^N}{N!} + \frac{x^{N+1}}{(N+1)!} + \frac{x^{N+2}}{(N+2)!} + \cdots \\
&= \frac{x^N}{N!}\left[1 + \frac{x}{N+1} + \frac{x^2}{(N+2)(N+1)} + \cdots\right] \\
&\leq \frac{x^N}{N!}\left[1 + \frac{x}{N+1} + \frac{x^2}{(N+1)^2} + \cdots\right] \\
&= \frac{x^N}{N!} \cdot \frac{1}{1 - x/(N+1)}
\end{aligned}$$

provided $x < N + 1$. For $x = 2$, this simplifies to

$$E_N(2) \leq \frac{2^N}{N!} \cdot \frac{N+1}{N-1}$$

and we require this quantity to be smaller than 10^{-5}. Now $2^{11}/11! = 5.130\,671\,8 \times 10^{-5}$, while $2^{12}/12! = 8.551\,119\,7 \times 10^{-6}$. We must check the effect of the factor $\dfrac{N+1}{N-1} = \dfrac{13}{11} = 1.181\,818\,2$. Since $(1.181\,818\,2)(8.551\,119\,7) = 10.105\,869$, 12 terms are not quite sufficient. $N = 13$ terms are needed: $\dfrac{2^{13}}{13!} \cdot \dfrac{14}{12} = 1.534\,816\,4 \times 10^{-6}$.

We note that for $|x| \leq 1$ in Example 3 we obtain $E_N(1) \leq \dfrac{N+1}{N \cdot N!} < 10^{-5}$ for $N \geq 9$. For $|x| \leq 1/2$, just 7 terms are needed. The number of terms required increases rapidly with x. These ideas can be used as a basis for *range reduction*, so that the series would be used only for very small values of x. For example, we could use the 7 terms to obtain $e^{1/2}$ and then take

$$e^2 = \left(e^1\right)^2 = \left(e^{1/2}\right)^4$$

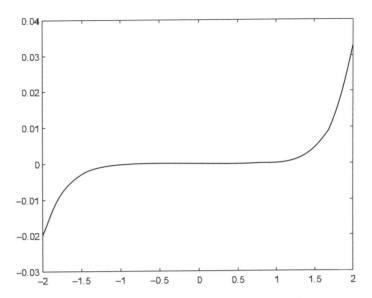

Figure 3.1   Series approximation error for e^x

to obtain exp 2. Greater care would be needed over the precision obtained from the series to allow for any loss of accuracy in squaring the result twice. The details are unimportant here, the point is that the series expansion can provide the basis for a good algorithm.

The magnitude of the error in any of these approximations to e^x increases rapidly with $|x|$, as can be seen from Figure 3.1.

The curve plotted is the error $e^x - \sum_{k=0}^{6} x^k / k!$. We see that the error remains very small throughout $[-1, 1]$ but rises sharply outside this interval indicating that more terms are needed there. The truncated exponential series is computed using the function m-file

```
function y=expn(x,n)
% Evaluates the first n terms of the exponential series
s=ones(size(x));
t=s;
for k=1:n-1
  t=t.*x/k;
  s=s+t;
end
y=s;
```

In the remaining examples of this section, we discover that substantial improvements are possible for the first two examples, too.

Example 4   Develop a series for $\ln((1+x)/(1-x))$, use it for the evaluation of $\ln 2$ with error less than 10^{-6}

We can use the series (3.7) and (3.8) to obtain

$$\ln\frac{1+x}{1-x} = \ln(1+x) - \ln(1-x)$$

$$= 2\left[x + \frac{x^3}{3} + \frac{x^5}{5} + \cdots\right] \tag{3.10}$$

Also $\dfrac{1+x}{1-x} = 2$ for $x = 1/3$ and so

$$\ln 2 = \frac{2}{3}\left[1 + \frac{(1/3)^2}{3} + \frac{(1/3)^4}{5} + \cdots\right] = \frac{2}{3}\sum_{k=0}^{\infty}\frac{1}{(2k+1)3^{2k}}$$

The truncation error incurred by using just the first N terms is then bounded by

$$\frac{2}{3}\cdot\frac{1}{2N+1}\cdot\frac{1}{9^N} < 10^{-6}$$

for $N \geq 6$ so that just 6 terms suffice. This compares very favourably with the 1 million that were required in Example 3. Using these 6 terms, we obtain the approximation

$$\ln 2 \approx \frac{2}{3}\sum_{k=0}^{5}\frac{1}{(2k+1)3^{2k}} = 0.693\,147\,07$$

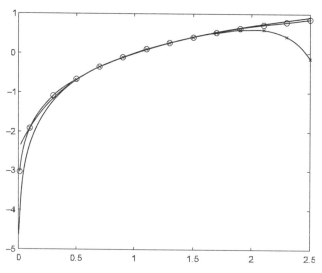

Figure 3.2   Series approximations to ln x

Recall the true value of $\ln 2 \approx 0.693\,147\,18$, so that the actual error is close to 10^{-7}, which is well within our tolerance.

The relative effectiveness of this approximation and the original power series (3.8) is illustrated in Figure 3.2. The true natural logarithm is plotted with a solid line, the sum of the first 6 terms of (3.8) with ×s and the first 6 terms of (3.10) with the os.

It is apparent that the new approximation reproduces the curve better over a wider range.

Example 5   Compute π using the identity $\arctan 1/\sqrt{3} = \pi/6$

In Example 2, we obtained the series expansion $\arctan x = x - \dfrac{x^3}{3} + \dfrac{x^5}{5} - \dfrac{x^7}{7} \cdots$ from which we have

$$\pi = 6 \left[3^{-1/2} - \frac{3^{-3/2}}{3} + \frac{3^{-5/2}}{5} \cdots \right]$$

$$= \frac{6}{\sqrt{3}} \left[1 - \frac{1}{3(3)} + \frac{1}{5(3)^2} \cdots \right] = 2\sqrt{3} \sum_{k=0}^{\infty} \frac{(-1)^k}{(2k+1)3^k}$$

The truncation error for this alternating series is bounded by the first term omitted. For single precision floating-point evaluation of π we would require this truncation error to be smaller than 2^{-23}, and so we seek N such that

$$\frac{2\sqrt{3}}{(2N+1)3^N} < 2^{-23}$$

This is satisfied for $N \geq 13$. This should be compared with the nearly 17 million terms needed to achieve the same accuracy in Example 2. Using these terms we obtain

$$\pi \approx 2\sqrt{3} \sum_{k=0}^{13} \frac{(-1)^k}{(2k+1)3^k} = 3.141\,592\,6$$

which is indeed a good approximation.

For IEEE double-precision (MATLAB) accuracy, the approximation in Example 5 requires just 30 terms. The 4 months on a fast computer for Example 2 would be reduced to about one-millionth of a second! Clearly the choice and design of computer algorithms can have a marked effect. The actual methods implemented on modern machines are much more efficient than these but the analysis of this section should give some idea of what can be achieved.

1   Write a MATLAB m-file to approximate the natural logarithm using the first 6 terms of (3.8). Use it to estimate $\ln 1.25$.

2   How many terms of the series (3.8) are needed to approximate $\ln 1.25$ with error smaller than 10^{-6}? Evaluate this approximation and verify that the error is indeed within the tolerance.

3   Use the fact that $\ln 10 = \ln 8 + \ln 1.25$, the result of Exercise 2, and an accurate value for $\ln 2$ to estimate $\ln 10$ with error smaller than 10^{-6}.

4   Determine the number of terms of the exponential series that are needed to obtain $e^{0.1}$ with error smaller than 10^{-10}. Evaluate the sum of these terms and verify that the desired accuracy is achieved.

5   Write a MATLAB m-file to compute the natural logarithm function using n terms of the approximation in Example 4. Graph the error in this approximation to $\ln x$ for $x \in (0, 2.5)$ using 8 terms.

6   The erf function, or 'error function', defined by

$$\text{erf}(x) = \frac{2}{\sqrt{\pi}} \int_0^x \exp\left(-t^2\right) dt$$

is an important function in statistics. Derive the series expansion

$$\text{erf}(x) = \frac{2}{\sqrt{\pi}} \sum_{k=0}^{\infty} \frac{(-1)^k x^{2k+1}}{(2k+1)k!}$$

Use the first 10 terms of this series to obtain a graph of this function over the interval $[0, 4]$. Compare this with the built-in erf function.

3.3   CORDIC algorithms

In this section, we concentrate on a topic which is in many ways a modern equivalent of interpolation in tables of logarithmic or trigonometric functions. Interpolation methods are still needed for special functions, and as a basis for numerical integration and the solution of differential equations. These aspects will be discussed in Chapters 4–6. The elementary functions such as ln, exp, sin and cos *are* available at the touch of a button on your calculator or computer.

Perhaps the most surprising aspect of this is that all these functions are computed on most calculators and PCs by minor variations of the same algorithm. The so-called *CORDIC* (*CO*ordinate *R*otation *DI*gital *C*omputer) algorithms were first developed by Volder for solving trigonometric problems in in-flight navigational computers. (In mathematics, it is usual that methods or theorems bear the names of their discoverers, and so perhaps these algorithms should be known as Volder methods. The use of the acronym CORDIC is testament to their having been discovered by an engineer.)

In this section, we shall concern ourselves only with binary versions of the CORDIC algorithms. Hand calculators typically use decimal-based versions but the extra detail of the decimal form of CORDIC algorithms does not enhance the understanding of the underlying ideas. To get a basic idea, we begin by observing that the long multiplication algorithm for binary integers is particularly simple, consisting solely of shifts and additions.

For example consider the multiplication 73×47. The decimal calculation requires 4 single-digit multiplications (each yielding a 2-digit result) with 2 carries and then a further addition with another carry. As a binary operation

$$73 \times 47 =$$

$$
\begin{array}{r}
1001001 \\
\times\ 101111 \\
\hline
1001001 \\
10010010 \\
100100100 \\
1001001000 \\
100100100000 \\
\hline
110101100111
\end{array}
$$

Each of these terms is just
1001001 shifted the
appropriate number of places.

The principal objective of the CORDIC algorithm is to achieve a similar level of simplicity for the elementary functions. All the CORDIC algorithms are based on an ingenious decomposition of the argument and/or the required answer in terms of simple constants, so that only additions and exponent shifts are needed. The following theorem lies at the heart of the matter.

Theorem 1 *Suppose that the numbers* σ_k $(k = 0, 1, \ldots, n)$ *are positive, decreasing and that*

$$\sigma_k \leq \sum_{j=k+1}^{n} \sigma_j + \sigma_n \tag{3.11}$$

Suppose too that

$$|r| \leq \sum_{j=0}^{n} \sigma_j + \sigma_n \tag{3.12}$$

Then the sequence defined by $s_0 = 0$ *and*

$$s_{k+1} = s_k + \delta_k \sigma_k \qquad (k = 0, 1, \ldots, n)$$

where $\delta_k = \mathrm{sgn}(r - s_k)$ *satisfies, for each* k,

$$|r - s_k| \leq \sum_{j=k}^{n} \sigma_j + \sigma_n \tag{3.13}$$

In particular, s_{n+1} *approximates* r *with error bounded by* σ_n, *that is,*

$$|r - s_{n+1}| \leq \sigma_n$$

> **Remark 1** Note that sgn is the usual signum function defined by
> $$sgn(x) = \begin{cases} 1 & \text{if } x \geq 0 \\ -1 & \text{if } x < 0 \end{cases}$$

Proof The proof is by induction. First, we have

$$|r - s_0| = |r| \leq \sum_{j=0}^{n} \sigma_j + \sigma_n$$

so that (3.13) holds for $k = 0$.

Now, assuming the result holds for some value of k, and noting that δ_k is chosen to have the same sign as $|r - s_k|$, we obtain

$$|r - s_{k+1}| = |r - s_k - \delta_k \sigma_k| = ||r - s_k| - \sigma_k|$$

Then, by (3.11) and the induction hypothesis, we deduce that

$$-\left(\sum_{j=k+1}^{n} \sigma_j + \sigma_n \right) \leq -\sigma_k \leq |r - s_k| - \sigma_k$$

$$\leq \left(\sum_{j=k}^{n} \sigma_j + \sigma_n \right) - \sigma_k = \sum_{j=k+1}^{n} \sigma_j + \sigma_n$$

as required. ∎

Theorem 1 shows that, if we choose a positive decreasing sequence $\sigma_0, \sigma_1, \ldots, \sigma_n$ to satisfy (3.11), then any number in the interval $[-E, E]$ where $E = \sum_{j=0}^{n} \sigma_j + \sigma_n$ can be written in the form $\pm\sigma_0 \pm \sigma_1 \pm \ldots \pm \sigma_n$ with an error no worse than σ_n.

Example 6 The values $\sigma_k = 2^{-k}$ satisfy the conditions of Theorem 1, and so for any $r \in [-2, 2]$, we can write

$$r \approx \pm 1 \pm \frac{1}{2} \pm \frac{1}{4} \pm \cdots \pm \frac{1}{2^n}$$

with error less than 2^{-n}. Find this decomposition of 1.2345 using $n = 5$

We set $s_0 = 0$ and choose $\delta_k = sgn(r - s_k)$. We get

$\delta_0 = +1$	$s_1 = 0 + 1(2^0) = 1$	$|r - s_1| = 0.2345$	$< \sigma_0 = 1$
$\delta_1 = +1$	$s_2 = 1 + 1(2^{-1}) = 1.5$	$|r - s_2| = 0.2655$	$< \sigma_1 = 1/2$
$\delta_2 = -1$	$s_3 = 1.5 - 1(2^{-2}) = 1.25$	$|r - s_3| = 0.0155$	$< \sigma_2 = 1/4$
$\delta_3 = -1$	$s_4 = 1.25 - 1(2^{-3}) = 1.125$	$|r - s_4| = 0.1095$	$< \sigma_3 = 1/8$
$\delta_4 = +1$	$s_5 = 1.125 + 1(2^{-4}) = 1.1875$	$|r - s_5| = 0.047$	$< \sigma_4 = 1/16$
$\delta_5 = +1$	$s_6 = 1.1875 + 1(2^{-5}) = 1.21875$	$|r - s_6| = 0.01575$	$< \sigma_5 = 1/32$

Clearly, this process could be continued to any desired accuracy. It is important to observe that the convergence is not monotone – even the absolute values of the errors can increase in individual steps as we see in steps 2 and 4 of Example 6.

The following general algorithm, with appropriate choices for the parameters m and σ_k, yields approximations to a wide class of elementary functions.

Algorithm 1 General CORDIC Algorithm

Inputs Starting values x_0, y_0, z_0
Parameter $m = -1$, 0, or $+1$, and corresponding sequence (σ_k)
Mode: either *rotation mode* or *vectoring mode*

Compute three sequences:

$$x_{k+1} = x_k - m\delta_k y_k 2^{-k} \tag{3.14}$$

$$y_{k+1} = y_k + \delta_k x_k 2^{-k} \tag{3.15}$$

$$z_{k+1} = z_k - \delta_k \sigma_k \tag{3.16}$$

where

$$\delta_k = \begin{cases} \operatorname{sgn}(z_k) & \text{for rotation mode} \\ -\operatorname{sgn}(y_k) & \text{for vectoring mode} \end{cases} \tag{3.17}$$

The value of m depends on the operation to be performed: $m = 0$ for arithmetic operations, $m = 1$ for trigonometric functions and $m = -1$ for hyperbolic functions. Details of the choice of the various parameters and modes for particular functions are summarized in Tables 3.1–3.3 below.

The names of the 2 modes are a historical accident owing to the development of the algorithms for navigational purposes. They are not of great value in understanding the methods.

Example 7 CORDIC Division

For multiplication and division we use $m = 0$ and $\sigma_k = 2^{-k}$ which we have already seen satisfy (3.11). Division is performed using the vectoring mode: with $z_0 = 0$, we find that

$$|z_{n+1} - y_0/x_0| < 2^{-n}$$

provided that the quotient $y_0/x_0 \in [-2, 2]$. To see this, we note first that, since $m = 0$, $x_k = x_0$ for every k, the algorithm reduces to just the 2 equations

$$y_{k+1} = y_k + \delta_k x_0 2^{-k}$$

$$z_{k+1} = z_k - \delta_k 2^{-k}$$

with $\delta_k = -\operatorname{sgn}(y_k)$.

In this case therefore (3.15) represents the decomposition

$$y_0 \approx -\sum_{k=0}^{n} \delta_k x_0 2^{-k}$$

or, equivalently,

$$\frac{y_0}{z_0} \approx -\sum_{k=0}^{n} \delta_k 2^{-k}$$

with error less than 2^{-n}. However, from (3.16) we see that $z_{n+1} = -\sum_{k=0}^{n} \delta_k 2^{-k}$.

As a specific example, consider the division $1.2/2.3$. Then $x_0 = 2.3$ and

$y_0 = 1.2$	$z_0 = 0$	$\delta_0 = -\text{sgn}(1.2) = -1$
$y_1 = 1.2 - (2.3)2^0 = -1.1$	$z_1 = 0 - (-1)2^0 = 1$	$\delta_1 = -\text{sgn}(-1.1) = +1$
$y_2 = -1.1 + (2.3)2^{-1} = 0.05$	$z_2 = 1 - (1)2^{-1} = 0.5$	$\delta_2 = -1$
$y_3 = 0.05 - (2.3)2^{-2} = -0.525$	$z_3 = 0.5 + 2^{-2} = 0.75$	$\delta_3 = +1$
$y_4 = -0.525 + (2.3)2^{-3} = -0.2375$	$z_4 = 0.75 - 2^{-3} = 0.625$	$\delta_4 = +1$
$y_5 = -0.09375$	$z_5 = 0.5625$	$\delta_5 = +1$
$y_6 = -0.021875$	$z_6 = 0.53125$	$\vdots$

The values of y_k are being driven towards 0 by the choice of the signs, and z_k is approaching $1.2/2.3 = 0.52174$ to 5 decimals. The error in z_6 is about 0.01 which is indeed smaller than $2^{-5} = 0.03125$.

We see that the most complicated operation involved here is the calculation of y_k, which involves no more than a binary shift of x_0 and the addition or subtraction of this quantity from the previous value. The algorithm entails no more than shifts and adds, which was the objective behind the development of CORDIC algorithms.

Multiplication can be performed using the same CORDIC algorithm in rotation mode. Setting $y_0 = 0$, we obtain $y_{n+1} \approx x_0 z_0$ with error bounded by $x_0 2^{-n}$. The details are left to the exercises.

Thus far, we have paid little attention to the condition (3.11) of Theorem 1, which for multiplication and division states that the decomposition works for

$$|r| \leq \sum_{k=0}^{n} 2^{-k} + 2^{-n} = 2$$

For division, this imposes the requirement that $|y_0/x_0| \leq 2$ which is automatically satisfied for (the mantissas of) normalized binary floating-point numbers which lie in $[1, 2)$. Similarly for floating-point multiplication, both operands are within the appropriate range. With n chosen appropriately, and with $x_0 \in [1, 2)$ the error bound for multiplication also guarantees the correct *relative* precision in the product.

The use of CORDIC algorithms for multiplication and division is summarized in Table 3.1. Here, as in Tables 3.2 and 3.3 for trigonometric and hyperbolic functions

Table 3.1 CORDIC algorithms for arithmetic operations

$m = 0$, $\sigma_k = 2^{-k}$ for $k = 0, 1, \ldots, n$

Function	Mode	Initial values	Output	Error bound	Useful domain
$*$	R	$y_0 = 0$	$y_{n+1} \approx x_0 z_0$	$\lvert x_0 \rvert 2^{-n}$	$\lvert x_0 \rvert, \lvert z_0 \rvert \leq 2$
$/$	V	$z_0 = 0$	$z_{n+1} \approx y_0 / x_0$	2^{-n}	$\lvert y_0 / x_0 \rvert \leq 2$

(pp. 68 and 72), R and V represent the Rotation and Vectoring modes, respectively. In each of the tables, a 'useful domain' is quoted for each operation. This is not necessarily the complete domain of convergence for the algorithm but it indicates a useful practical domain for which convergence can be established.

Example 8   CORDIC trigonometric functions

For the trigonometric functions, we use $m = 1$ and take

$$\sigma_k = \arctan 2^{-k} \qquad (k = 0, 1, \ldots, n)$$

The fact that this sequence satisfies (3.11) can be established by an application of the Mean Value Theorem.

The rotation mode provides a technique for computing $\sin \theta$ and $\cos \theta$ by setting $z_0 = \theta$ and decomposing this as $\sum \delta_k \sigma_k$. Writing $s_k = \theta - z_k$, it follows from (3.16) that

$$s_{k+1} = s_k + \delta_k \sigma_k$$

Hence, using the facts that sine and cosine are odd and even functions, respectively, we have

$$\begin{aligned}
\cos(s_{k+1}) &= \cos(s_k + \delta_k \sigma_k) \\
&= \cos(s_k) \cos(\delta_k \sigma_k) - \sin(s_k) \sin(\delta_k \sigma_k) \\
&= \cos(s_k) \cos(\sigma_k) - \delta_k \sin(s_k) \sin(\sigma_k)
\end{aligned}$$

and, similarly,

$$\sin(s_{k+1}) = \sin(s_k) \cos(\sigma_k) + \delta_k \cos(s_k) \sin(\sigma_k)$$

Dividing both these equations by $\cos(\sigma_k)$ and observing that $\tan(\sigma_k) = 2^{-k}$, we now obtain

$$\frac{\cos(s_{k+1})}{\cos(\sigma_k)} = \cos(s_k) - \delta_k 2^{-k} \sin(s_k) \qquad (3.18)$$

$$\frac{\sin(s_{k+1})}{\cos(\sigma_k)} = \sin(s_k) + \delta_k 2^{-k} \cos(s_k) \qquad (3.19)$$

From Theorem 1 it follows that $|z_{n+1}| = |s_{n+1} - \theta| \le \sigma_n$. Apart from the divisor $\cos(\sigma_k)$, (3.18) and (3.19) resemble (3.14) and (3.15) with $m = 1$, $x_k = \cos(s_k)$, and $y_k = \sin(s_k)$. The factors $\cos(\sigma_k)$ and their product

$$K_T = \prod_{k=0}^{n} \cos(\sigma_k)$$

can be precomputed. The initial values of x_k and y_k can be premultiplied by this constant K_T. That is, we set

$$x_0 = K_T \cos(s_0) = K_I \cos(0) = K_T$$
$$y_0 = K_T \sin(s_0) = K_T \sin(0) = 0$$

The effect is that after the $n + 1$ steps are completed, we have

$$x_{n+1} \approx \cos\theta, \quad y_{n+1} \approx \sin\theta$$

each with error less than 2^{-n}.

To illustrate this algorithm we shall compute $\sin(1)$ and $\cos(1)$ using just 4 steps ($n = 3$). In this case we use

$$\sigma_0 = \arctan 1 = 0.7854$$
$$\sigma_1 = \arctan(1/2) = 0.4636$$
$$\sigma_2 = \arctan(1/4) = 0.2450$$
$$\sigma_3 = \arctan(1/8) = 0.1244$$

and

$$K_T = \cos(\sigma_0)\cos(\sigma_1)\cos(\sigma_2)\cos(\sigma_3) = 0.6088$$

Then, with $\delta_k = \text{sgn}(z_k), m = 1$ we get, using (3.14)–(3.16)

$x_0 = 0.6088$	$y_0 = 0$	$z_0 = 1$	$\delta_0 = +1$
$x_1 = 0.6088$	$y_1 = 0.6088$	$z_1 = 0.2146$	$\delta_1 = +1$
$x_2 = 0.3044$	$y_2 = 0.9132$	$z_2 = -0.2490$	$\delta_2 = -1$
$x_3 = 0.5327$	$y_3 = 0.8371$	$z_3 = -0.0040$	$\delta_3 = -1$
$x_4 = 0.6373$	$y_4 = 0.7705$	$z_4 = 0.1204$	$\delta_4 = +1$

from which we deduce that $\cos 1 \approx 0.6373$ and $\sin 1 \approx 0.7705$, each with error less than $2^{-3} = 0.125$. (The true values are 0.5403 and 0.8415.)

Note that, as with the division algorithm, the errors are not necessarily reduced at each iteration. The compensating advantage is that we know *in advance* the exact number of steps that are needed to achieve a specified accuracy. In the case of the trigonometric functions, we should also note that it is not until the completion of the predetermined number of steps that we really have approximations to $\cos\theta$ and $\sin\theta$ because of the initial scaling by K_T which depends on the number of steps to be used.

| Program | **MATLAB m-file for sin and cos using CORDIC algorithm with 40 steps** |

```
function y=Cordictrig(z)
% Computes cos and sin of z using 40 steps of CORDIC algorithm
% y(1), y(2) are approximations of cos(z), sin(z) respectively
s=2.^-(0:39); sig=atan(s);
KT=prod(cos(sig));
x1=KT; y1=0;
for k=1:40
  x0=x1; y0=y1;
  if z>=0, del=1; else del=-1; end
  x1=x0-del*y0*s(k);
  y1=y0+del*x0*s(k);
  z=z-del*sig(k);
end
y(1)=x1; y(2)=y1;
```

The results from this m-file should be accurate to within 2^{-39}. As an example, the command

» cs=cordictrig(1)

yields the result

cs =
0.540302305868555 0.841470984807631

which each have errors smaller than 2^{-41} as can be verified using

» log2(abs([cos(1)-cs(1),sin(1)-cs(2)]))
ans =
 −41.1327212602903   −41.7753953184615

The vectoring mode of the trigonometric CORDIC provides algorithms for both the inverse tangent function and for the magnitude of a 2-dimensional vector. Specifically, with $z_0 = 0$, we can use it to compute

$$z_{n+1} \approx \arctan(y_0/x_0)$$

and

$$x_{n+1} \approx \sqrt{x_0^2 + y_0^2}/K_T$$

If the initial values x_0, y_0 are scaled by K_T, the arctangent value is unchanged and the square root becomes just $\sqrt{x_0^2 + y_0^2}$, which is to say we have precisely the appropriate output for conversion between Cartesian and polar coordinates in the plane.

> **Remark 2** It may appear from the program above that this algorithm requires knowledge of both arctan and cos in order to compute these functions. However, we should note that in a hardware implementation only those special values σ_k and $\cos \sigma_k$ are needed and these would be precomputed and stored on the chip.

As with the arithmetic functions, we must consider the intervals of convergence for these CORDIC trigonometric functions. Since the angle θ is decomposed as $\sum \delta_k \sigma_k$, this algorithm will be applicable for any $|\theta| \leq \sum \sigma_k + \sigma_n \approx 1.74$ which is greater than $\pi/2$ so that range reduction could be implemented to allow the CORDIC algorithm to compute any value of these functions. Essentially this involves subtracting the appropriate integer multiple of 2π and then adjusting the answers for the correct quadrant. For arguments outside a moderate interval, this range reduction is not trivial. We do not concern ourselves with the details here.

The complete domain of the arctangent can be covered since, using $x_0, y_0 \in [-2, 2]$, the range of values of y_0/x_0 spans the whole real line. For the geometric coordinate transformation, this range remains sufficient since we can scale both coordinates by an appropriate binary exponent. As before we summarize the trigonometric CORDIC algorithms in a table (Table 3.2).

Table 3.2 CORDIC algorithms for trigonometric functions

$m = 1$, $\sigma_k = \arctan 2^{-k}$ for $k = 0, 1, \ldots, n$, $K_T = \prod_{k=0}^{n} \cos(\sigma_k)$				
Function	*Mode*	*Initial values*	*Output*	*Useful domain*
cos, sin	R	$x_0 = K_T$ $y_0 = 0$	$x_{n+1} \approx \cos z_0$ $y_{n+1} \approx \sin z_0$	$\|x_0\|, \|z_0\| \leq \pi/2$
arctan	V	$z_0 = 0$	$z_{n+1} \approx \arctan(y_0/x_0)$	$\|x_0\|, \|y_0\| \leq 2$
$\|\cdot\|$	V	$x_0 = xK_T$ $y_0 = yK_T$	$x_{n+1} \approx \sqrt{x^2 + y^2}$	$\|x\|, \|y\| \leq 2$

Example 9   Hyperbolic, exponential and logarithmic functions

The CORDIC algorithms for these functions are very similar to those for the trigonometric functions. This time we take $m = -1$ and $\sigma_k = \tanh^{-1} 2^{-k}$ for $k \geq 1$. Equations similar to (3.18) and (3.19) can be derived from the corresponding identities for the hyperbolic functions. There is one very important difference, however. The quantities σ_k just defined *do not* satisfy condition (3.11). It can be shown, however, that, if the steps for $k = 4, 13, 40, \ldots$, or, in general, $k = (3^j - 1)/2$, are repeated then all the conditions of Theorem 1 are satisfied. Corresponding to the quantity K_T used for the trigonometric functions, this time we take

$$K_H = \prod \cosh \sigma_k$$

where the product includes repetitions of the appropriate factors.

In the rotation mode if $x_1 = K_H$ and $y_1 = 0$ then

$$x_{n+1} \approx \cosh z_1$$
$$y_{n+1} \approx \sinh z_1$$

each with error smaller than 2^{-n}. From these we can obtain the exponential function since

$$e^{z_1} = \cosh z_1 + \sinh z_1$$

With $z_1 = 0$, the vectoring mode can be used to obtain

$$z_{n+1} \approx \tanh^{-1}(y_1/x_1)$$
$$= \frac{1}{2} \ln w \tag{3.20}$$

if $x_1 = w + 1$, $y_1 = w - 1$. Also

$$x_{n+1} \approx \frac{\sqrt{x_1^2 - y_1^2}}{K_H} = \frac{\sqrt{w}}{K_H} \tag{3.21}$$

if $x_1 = w + 1/4$, $y_1 = w - 1/4$.

We illustrate these algorithms with the estimation of $e^{0.2}$ using $n = 5$. The basic algorithm can be simplified somewhat for the exponential function. Equations (3.14) and (3.15) with $m = -1$ become

$$x_{k+1} = x_k + \delta_k y_k 2^{-k}$$
$$y_{k+1} = y_k + \delta_k x_k 2^{-k}$$

and writing $u_k = x_k + y_k$ we obtain the single equation

$$u_{k+1} = u_k + \delta_k u_k 2^{-k} \tag{3.22}$$

Now using $n = 5$, which corresponds to 6 steps, and allowing for the repetition for $k = 4$, we use

$$\sigma_1 = \tanh^{-1} 2^{-1} = 0.5493, \ \sigma_2 = 0.2554,$$
$$\sigma_3 = 0.1257, \ \sigma_4 = 0.0626, \ \sigma_5 = 0.0626$$

from which we have

$$K_H = \cosh(0.5493) \cosh(0.2554) \cosh(0.1257) \cosh^2(0.0626)$$
$$= 1.2067$$

Then we obtain

$$u_1 = 1.2067 \quad z_1 = 0.2000 \quad \delta_1 = +1$$
$$u_2 = 1.8101 \quad z_2 = -0.3493 \quad \delta_2 = -1$$
$$u_3 = 1.3576 \quad z_3 = -0.0939 \quad \delta_3 = -1$$
$$u_4 = 1.1879 \quad z_4 = 0.0318 \quad \delta_4 = +1$$
$$u_5 = 1.2621 \quad z_5 = -0.0308 \quad \delta_5 = -1$$
$$u_6 = 1.1832 \quad z_6 = 0.0318 \quad \delta_6 = +1$$

Hence, we obtain $e^{0.2} \approx u_7 = 1.1832 + (1.1832)2^{-5} = 1.2202$, which should be compared with the true value 1.2214.

In the following program, the set of steps that are repeated includes $k = 1$. This has the benefit of increasing the range of applicability of the algorithm. This program implements the vectoring mode. With careful choice of the inputs, the outputs yield not only the inverse hyperbolic tangent but also the natural logarithm and square-root functions.

Program

MATLAB m-file for $\tanh^{-1}$ using CORDIC algorithm with 40 steps plus repetitions

```
function out=cordichypv(x,y)
% Cordic algorithm for hyperbolic functions
% Vectoring mode, 40 steps plus repetitions for 1,4,13,40
v=[1,1:4,4:13,13:40,40];
% Note this takes account of the repetitions
s=2.^-v;
sig=atanh(s);
KH=prod(cosh(sig));
z=0; x1=x; y1=y;
for k=1:44
   x0=x1; y0=y1;
   if y0>=0, del=-1; else del=1; end
   x1=x0+del*y0*s(k);
   y1=y0+del*x0*s(k);
   z=z-del*sig(k);
end
out=[x1,y1,z];
```

For example, the command

» v=cordichypv(2,1)

yields the output vector

v = 1.24223904144032, 9.7878783548196e-014, 0.549306144333976

the third element of which is the approximation to $\tanh^{-1}(1/2)$ which has true value 0.549306144334055. The error is about 10^{-13}.

The second element of the output merely confirms that the CORDIC iterations have forced y_k to approach 0. The first element is the final x_{n+1} value which should be close to $\sqrt{2^2 - 1^2}/K_H = \sqrt{3}/1.39429751423739 = 1.242\,239$ to 6 decimals.

Equations (3.20) and (3.21) show how to modify the initial values to obtain other desired outputs. For example, to get ln 2, we must choose $x_1 = 2 + 1 = 3$, $y_1 = 2 - 1 = 1$. This will result in $z_{n+1} \approx (1/2) \ln 3$ and so the third element of the output vector must be doubled:

» v=cordichypv(3,1);

» ln2=2*v(3)

ln2 =

    0.69314718056046

which should be compared with the true value 0.693147180559945.

The convergence domain for the hyperbolic functions with the repetitions used is $\sum \sigma_k \approx 1.74$ so that $\cosh z$, $\sinh z$ and $\exp z$ may be computed for $|z| \leq 1.74$. This range can be extended in a variety of ways. One convenient method is to write larger arguments in the form

$$z = z_1 + p \ln 2$$

where p is an integer chosen so that $z_1 \in [0, \ln 2)$. Then z_1 is in the range of applicability and we can then use

$$e^z = e^{p \ln 2} e^{z_1} = 2^p e^{z_1}$$

which will be the normalized binary floating-point representation. The error in the value of e^{z_1} will be 2^{-n+1}, where n is the number of steps used. It is therefore easy to determine in advance the number of steps needed for any particular floating-point format. This number of steps remains fixed for all arguments.

The CORDIC algorithms for the hyperbolic functions are summarized in Table 3.3.

It is necessary here to give a word of warning about the possibility of meaningless computation. As an illustration, suppose that a hypothetical calculator works to 7 decimal digits so that a number, A, is represented as $a \times 10^{\alpha}$ with $1 \leq a < 10$. The difference between A and the next representable number is then $10^{\alpha-6}$ which is certainly greater than 2π whenever $\alpha \geq 7$. To try to give a specific value to, say, $\cos A$ is then plainly meaningless since more than one complete period of the cosine function would share the same representation. Nonetheless most computers and calculators will attribute a specific value to $\cos A$, which emphasizes the point that *any* output from a computer or calculator should be treated with suspicion until the situation has been analyzed carefully.

Before leaving the subject of approximate evaluation of functions, it should be stressed that, although many practical algorithms are based on the ideas presented here, these are by no means the only ones available. The different routines used in various computers provide ample testimony to the variety and blend of approaches

Table 3.3   CORDIC algorithms for hyperbolic functions

$m = -1$, $\sigma_k = \tanh^{-1} 2^{-k}$ for $k = 1, 2, \ldots, n$, $K_H = \prod \cosh(\sigma_k)$ with repetitions for $k = 1, 4, 13, 40, \ldots$

Function	Mode	Initial values	Output	Useful domain
cosh sinh exp	R	$x_0 = K_H$ $y_0 = 0$	$x_{n+1} \approx \cosh z_0$ $y_{n+1} \approx \sinh z_0$ $x_{n+1} + y_{n+1} \approx e^{z_0}$	$\lvert x_0 \rvert, \lvert y_0 \rvert \le 1.7$
$\tanh^{-1}$	V	$z_1 = 0$	$z_{n+1} \approx \tanh^{-1}(y_1/x_1)$	$\lvert y_1 \rvert < \lvert x_1 \rvert$ $\lvert y_1 \rvert \le 2$
$\ln w$	V	$x_1 = w + 1$ $y_1 = w - 1$	$z_{n+1} \approx \frac{1}{2} \ln w$	$1/2 \le w \le 2$
$\sqrt{w}$	V	$x_1 = K_H(w + 1/4)$ $y_1 = K_H(w - 1/4)$	$x_{n+1} \approx \sqrt{w}$	$1 \le w \le 4$

which may be useful in different circumstances. The interpolation-based methods of Chapter 4 also play an important role, as do others that are beyond our present objectives.

Exercises: Section 3.3

1 Approximate 0.12345 in the form $\pm 1 \pm 1/2 \pm \cdots \pm 1/32$. Verify that the error satisfies the bound given in Theorem 1.

2 Show that $\sigma_k = 2^{-k}$ for $k = 0, 1, \ldots, n$ satisfies the condition (3.11) of Theorem 1.

3 Write a MATLAB m-file to obtain the CORDIC decomposition of a number $r \in [-2, 2]$ using $\sigma_k = 2^{-k}$ for $k = 0, 1, \ldots, 40$. Test it for $r = \pm 0.12345$ and ± 1.2345 and verify that the error bounds satisfy (3.13).

4 Use the CORDIC multiplication algorithm to compute 1.23×1.12 with error less than 2^{-7}.

5 Show that $\sigma_k = \arctan 2^{-k}$ for $k = 0, 1, \ldots, n$ satisfies condition (3.11) of Theorem 1. Use the CORDIC algorithm to compute $\sin(0.5)$ and $\cos(0.5)$ with $n = 5$.

6 Write a MATLAB m-file to compute the vectoring mode of the trigonometric CORDIC algorithm using 40 steps. Use it to convert the Cartesian coordinates $(2, 1)$ to plane polar coordinates.

7 Derive a simplified CORDIC scheme for evaluating the function e^{-z}. Use this algorithm with $n = 6$ to approximate $e^{-0.25}$.

8 Use the CORDIC algorithm to approximate $\ln 1.5$ using $n = 6$.

9 Write an m-file for the rotation mode of the CORDIC algorithm for hyperbolic functions using $n = 40$. (Don't forget the repeated steps!) Use this to evaluate $\sinh x$ for $x = -1.5 : 0.1 : 1.5$. Graph the error function for these values and verify that the errors are appropriately bounded.

3.4 MATLAB functions

MATLAB has algorithms built in for all the standard 'elementary' mathematical functions, such as the trigonometric functions and their inverses, the exponential function, the natural logarithm and the hyperbolic functions and their inverses.

Their names are usually the expected ones with the convention that the inverse functions are prefixed with the letter a so that atan is MATLAB's arctan function. The syntax is also as you would expect. Some of the basic ones are included in Table 3.4. All the standard MATLAB functions can be applied elementwise to vectors and matrices – a fact which is particularly helpful for graphics.

Table 3.4   MATLAB elementary functions

Mathematical notation	MATLAB	Mathematical notation	MATLAB
$\sin x$	sin(x)	$e^x = \exp(x)$	exp(x)
$\cos x$	cos(x)	$\ln x$	log(x)
$\tan x$	tan(x)	$\log_{10} x$	log10(x)
$\arctan x$	atan(x)	$\log_2 x$	log2(x)
$\arcsin x$	asin(x)	$\cosh x$	cosh(x)
$\sqrt{x}$	sqrt(x)	$\sinh x$	sinh(x)
$|x|$	abs(x)	$\tanh^{-1} x$	atanh(x)

Modern PCs can evaluate most of these in hardware using algorithms similar to those described in this chapter. However, MATLAB *does not* use these hardware functions because its results should be the same independent of the hardware platform being used. (Most UNIX workstations do not have the ability to compute these in hardware on their RISC (Reduced Instruction Set Computer) chips.) MATLAB employs very efficient software code for these functions.

In addition to these 'elementary' functions, MATLAB has built-in m-files for many other 'special' functions. These include the erf function (which was introduced in the Exercises to Section 3.2), the beta and gamma functions, the Bessel functions and many others. Again the syntax is much as would be expected for these different functions. You can check the details in your MATLAB documentation when you need to use any of these functions.

4 Interpolation

Aims and objectives

We have considered the evaluation of elementary functions. In this chapter, we turn to the approximate evaluation of a function which is known only by its values at a set of data points. *Interpolation* is the process of fitting a function of a particular nature (typically a polynomial, or a collection of polynomials) through the data. We begin with the general theory of polynomial interpolation, and conclude with using *splines* as our interpolating functions. As usual, we strike a balance between the mathematical theory of the methods under consideration and their practical implementation on a computer.

4.1 Introduction

In Chapter 3, we studied two basic approaches to the approximation of functions. It was mentioned there that series and the CORDIC algorithms are by no means the only ways of approximating values of functions. In this chapter, we introduce some of the alternatives.

We begin with *polynomial interpolation*, which is based on the idea of finding a polynomial which agrees exactly with some information that we have about the function under consideration. This may be in the form of values of the function at some set of points, or may include some values of derivatives of that function. Among the situations where polynomial interpolation may be useful are those where the only information we have is in this form. This would be the case if, for example, the data are the result of some physical experiment, or if we are given a table of values of the function.

Traditional polynomial interpolation is not always appropriate, although it gives a useful and instructive approach to approximation. In many circumstances, other methods will prove preferable, and it is helpful to draw a distinction between the approximation of a smooth function whose tabulated values are known very accurately, and the situation where the data itself is subject to error. (Most practical problems have some aspects of both of these.) In the former situation, *polynomial approximation* may be applied with advantage over a wide interval, sometimes using polynomials of moderate or high degree. In the latter case, it is invariably desirable to use lower-degree polynomials over restricted ranges. One important technique is approximation by *cubic splines* in which the function is represented by different cubic polynomials in different intervals. This representation is smoothed by enforcing continuity of the first two derivatives throughout the range.

In this chapter we shall be concerned only with interpolation techniques – that is, the approximating function and the function being approximated share the same values at the data points. When data is subject to error (experimental error, for example) it is common to use approximation methods which force the approximating function to pass close to all the data points without necessarily going through them. One of the basic approximation methods is least squares approximation, which is discussed in Chapter 7.

Before discussing any of the techniques in detail, it is worth making a few introductory remarks about the reasons for choosing polynomials as the basic tools for approximating functions. There are two principal reasons. First, any continuous function can be approximated to any required accuracy by a polynomial. Secondly, polynomials are the *only* functions, which we can, theoretically at least, evaluate *exactly*.

The first of these statements is based on a famous theorem of Weierstrass.

Theorem 1 (Weierstrass) *Let f be a continuous function on the interval $[a, b]$. Given any $h > 0$, there exists a polynomial $p_{N(h)}$ of degree $N(h)$ such that*

$$\left| f(x) - p_{N(h)}(x) \right| < h$$

for all $x \in [a, b]$; therefore there exists a sequence of polynomials such that $\| f - p_n \|_\infty \to 0$ as $n \to \infty$.

The second reason for choosing polynomials also merits further comment. The validity of the statement is left for you to ponder, but it is worth spending a few moments on the question of the efficient evaluation of a polynomial. This is best achieved by using *Horner's rule*.

Horner's Rule   Efficient evaluation of a polynomial

Suppose we wish to evaluate

$$p(x) = a_n x^n + a_{n-1} x^{n-1} + \cdots + a_1 x + a_0$$

Horner's rule states that

$$p(x) = \{ \cdots [(a_n x + a_{n-1}) x + a_{n-2}] x + \cdots + a_1 \} x + a_0$$

This is easily implemented in MATLAB using the following code:

```
function p=horner(a,x)
% Horner's rule for evaluating the polynomial
% p(x)=a(1)+a(2)x+...+a(n-1)x^(n-2)+a(n)x^(n-1)
% NOTE the subscripts on the array are shifted
% because MATLAB arrays start with element 1
n=length(a);
p=a(n);
for k=n-1:-1:1
    p=p*x+a(k);
end
```

For example, the polynomial $x^2 + 3x + 2$ could be evaluated at $x = 2$ by

```
» horner([2,3,1],2)
```

which returns the expected value 12.

MATLAB's function polyval is essentially similar – except that it has the coefficient vector in the reverse order.

Considering this as a piece of hand calculation, we see a significant saving of effort. In the m-file above each of the $n - 1$ steps entails a multiplication and an addition, so that for a degree $n - 1$ polynomial, a total of $n - 1$ additions and $n - 1$ multiplications are needed. Direct computation requires much more than this. Evaluation of $a_k x^k$ requires k multiplications, and therefore the complete operation would need

$$(n - 1) + (n - 2) + \cdots + 2 + 1 = \frac{n(n - 1)}{2}$$

multiplications and $n - 1$ additions. Horner's rule is also numerically more stable, though this is less easy to illustrate without resorting to highly pathological examples.

Example 1   **Use Horner's rule to evaluate**

$$p(x) = 6x^5 + 5x^4 + 4x^3 + 3x^2 + 2x + 1$$

for $x = 0.1234$

We get

$$p(x) = (\{[[(6x + 5)x + 4]x + 3\}x + 2)x + 1$$
$$= ((((6(0.1234) + 5)(0.1234) + 4)(0.1234) + 3)(0.1234) + 2)(0.1234) + 1$$
$$= 1.3013301$$

To 6 decimals the sequence of values generated by the m-file above is:

6, 5.7404, 4.708365, 3.581012, 2.441897, 1.301330

We see that polynomials offer us both ease of evaluation and arbitrary accuracy of approximation. These properties make them a natural starting point for interpolation.

Exercises:
Section 4.1

1 Use Horner's rule to evaluate $\sum_{k=0}^{10} kx^k$ for $x = 1/2$.

2 Repeat Exercise 1 using the m-file horner.m and using MATLAB's polyval command.

3 Modify horner.m for MATLAB vector inputs. Use it to evaluate the polynomial in Exercise 1 for $x = -1 : 0.1 : 2$. Use the resulting data to plot a graph of this polynomial.

4.2 Lagrange interpolation

The basic idea of polynomial interpolation is that we find a polynomial which agrees with the data from the function f of interest. The *Lagrange interpolation polynomial* has the property that it takes the same values as f at a finite set of distinct points.

Before discussing the details of this, it should be pointed out that you are probably familiar with at least one form of interpolation polynomial. The first $N + 1$ terms of the Taylor expansion of f about a point x_0 form a polynomial, the *Taylor polynomial*, of degree N, namely

$$p(x) = f(x_0) + (x - x_0)f'(x_0) + \cdots + \frac{(x - x_0)^N}{N!} f^{(N)}(x_0)$$

which satisfies the *interpolation conditions*

$$p^{(k)}(x_0) = f^{(k)}(x_0) \qquad (k = 0, 1, \ldots, N)$$

We also know from Taylor's theorem that the error in using $p(x)$ to approximate $f(x)$ is given by

$$f(x) - p(x) = \frac{(x - x_0)^{N+1}}{(N + 1)!} f^{(N+1)}(\xi)$$

for some ξ lying between x and x_0.

Why should we not just settle for this? First, in many data-fitting applications, we are only given function values and the corresponding derivative values are not available; secondly, this approach will usually provide good approximations only near the base point x_0.

The Taylor polynomial does, however, illustrate the general polynomial interpolation approach: we first find a formula for the polynomial (of minimum degree) which satisfies the required interpolation conditions, and then find an expression for the error of the resulting approximation. Of course, this error will not be an explicitly computable quantity since, if it were, then we could evaluate the function exactly in the first place.

To develop the theory of the Lagrange interpolation polynomial, suppose then that we are given the values of a function f at $N + 1$ distinct points, called *nodes*, $x_0, x_1, \ldots, x_N$. We wish to find the minimum degree polynomial p such that

$$p(x_k) = f(x_k) \qquad (k = 0, 1, \ldots, N) \tag{4.1}$$

Suppose that p has degree m. Then we can write

$$p(x) = a_m x^m + a_{m-1} x^{m-1} + \cdots + a_1 x + a_0 \tag{4.2}$$

where the coefficients $a_0, a_1, \ldots, a_m$ are, as yet, undetermined. Substituting (4.2) into (4.1) yields the following system of linear equations for these unknown coefficients:

$$\begin{aligned}
a_0 + a_1 x_0 + \cdots + a_{m-1} x_0^{m-1} + a_m x_0^m &= f(x_0) \\
a_0 + a_1 x_1 + \cdots + a_{m-1} x_1^{m-1} + a_m x_1^m &= f(x_1) \\
&\vdots \\
a_0 + a_1 x_N + \cdots + a_{m-1} x_N^{m-1} + a_m x_N^m &= f(x_N)
\end{aligned} \tag{4.3}$$

In general, such a system will have no solution if $m < N$, infinitely many solutions if $m > N$, and, provided the matrix A with elements $a_{ij} = x_{i-1}^{j-1}$ is nonsingular (which it is), a unique solution if $m = N$. We thus expect our interpolation polynomial to have degree N (or less if it turns out that $a_N = 0$). Although (4.3) provides us with a theoretical way of finding the coefficients, if the idea is to be useful, we require a more convenient route to finding the polynomial p satisfying (4.1).

Now IF (that's right, this is a 'big if') we can find polynomials l_j $(j = 0, 1, \ldots, N)$ of degree at most N such that

$$l_j(x_k) = \delta_{jk} = \begin{cases} 1 & \text{if } j = k \\ 0 & \text{if } j \neq k \end{cases} \tag{4.4}$$

then the polynomial p given by

$$p(x) = \sum_{j=0}^{N} f(x_j) l_j(x) \tag{4.5}$$

will have degree at most N and will satisfy the interpolation conditions (4.1).

Before obtaining these polynomials explicitly, we verify the claim that p defined by (4.5) satisfies the desired interpolation conditions. Now

$$p(x_k) = \sum_{j=0}^{N} f(x_j) l_j(x_k) = \sum_{j=0}^{N} f(x_j) \delta_{jk}$$
$$= f(x_k)$$

by (4.4).

We now have some incentive to find the l_j. The requirement that $l_j(x_k) = 0$ whenever $j \neq k$ means that $l_j(x)$ must have factors $(x - x_k)$ for each such k. There are N such factors so that

$$l_j(x) = c \prod_{k \neq j} (x - x_k) = c(x - x_0) \cdots (x - x_{j-1}) \cdot (x - x_{j+1}) \cdots (x - x_N)$$

has degree N and satisfies $l_j(x_k) = 0$ for $k \neq j$. It remains to choose the constant c so that $l_j(x_j) = 1$. This condition implies that

$$c = \frac{1}{\prod_{k \neq j} (x_j - x_k)}$$

so that

$$l_j(x) = \prod_{k \neq j} \frac{(x - x_k)}{(x_j - x_k)} = \frac{(x - x_0) \cdots (x - x_{j-1}) \cdot (x - x_{j+1}) \cdots (x - x_N)}{(x_j - x_0) \cdots (x_j - x_{j-1}) \cdot (x_j - x_{j+1}) \cdots (x_j - x_N)}. \quad (4.6)$$

These polynomials are called the *Lagrange basis polynomials* and the polynomial p given by (4.5) is the *Lagrange interpolation polynomial*.

We have already established the existence of this polynomial by finding it, although we had previously asserted this from consideration of the linear equations (4.3). These same considerations also demonstrate the uniqueness of the interpolation polynomial of degree at most N satisfying (4.1). This relies on the stated fact that the matrix A mentioned above is nonsingular. The proof of this, by showing that the *Vandermonde determinant*

$$V = \begin{vmatrix} 1 & x_0 & x_0^2 & \cdots & x_0^N \\ 1 & x_1 & x_1^2 & \cdots & x_1^N \\ & \cdots & & & \cdots \\ 1 & x_N & x_N^2 & \cdots & x_N^N \end{vmatrix}$$

is nonzero for distinct nodes $x_0, x_1, \ldots, x_N$, is left as an exercise. An alternative proof of uniqueness can be obtained from the Fundamental Theorem of Algebra since we cannot find two *distinct* polynomials p and q of degree N which have the same values at $N + 1$ distinct nodes. (This would imply that the difference $p - q$ would have $N + 1$ distinct roots.)

Example 2 **Find the Lagrange interpolation polynomial for the following data and use it to estimate $f(1.2)$**

x	1	1.5	2
$f(x)$	0.0000	0.4055	0.6931

With $x_0 = 1$, $x_1 = 1.5$, and $x_2 = 2$, we obtain the quadratic interpolation polynomial

$$p(x) = f(x_0)\frac{(x - x_1)(x - x_2)}{(x_0 - x_1)(x_0 - x_2)} + f(x_1)\frac{(x - x_0)(x - x_2)}{(x_1 - x_0)(x_1 - x_2)}$$
$$+ f(x_2)\frac{(x - x_0)(x - x_1)}{(x_2 - x_0)(x_2 - x_1)}$$
$$= 0.0000\frac{(x - 1.5)(x - 2)}{(1 - 1.5)(1 - 2)} + 0.4055\frac{(x - 1)(x - 2)}{(1.5 - 1)(1.5 - 2)}$$
$$+ 0.6931\frac{(x - 1)(x - 1.5)}{(2 - 1)(2 - 1.5)}$$

For $x = 1.2$, this yields

$$p(1.2) = 0.4055\frac{(1.2 - 1)(1.2 - 2)}{(1.5 - 1)(1.5 - 2)} + 0.6931\frac{(1.2 - 1)(1.2 - 1.5)}{(2 - 1)(2 - 1.5)}$$
$$= 0.176348$$

from which we obtain the approximation $f(1.2) \approx 0.1763$ to 4 decimal places.

Since the original data is given to just four decimals, giving any greater accuracy in this answer would be unjustified.

As with the Taylor polynomial, we wish to find an expression for the error in this Lagrange interpolation process. The proof of the error formula is accomplished with the repeated use of Rolle's theorem which states that between any two zeros of a differentiable function there must be a point at which the derivative is zero. (For a more precise statement of this theorem consult your calculus or elementary analysis text.)

Theorem 2 *Error in Lagrange interpolation*
Suppose the function f is $N + 1$ times continuously differentiable on the interval $[a, b]$ which contains the distinct nodes $x_0, x_1, \ldots, x_N$. Let p be the Lagrange interpolation polynomial given by equations (4.5) and (4.6). Then, for any $x \in [a, b]$,

$$f(x) - p(x) = \frac{(x - x_0)(x - x_1) \cdots (x - x_N)}{(N + 1)!} f^{(N+1)}(\xi) \tag{4.7}$$

for some $\xi \in [a, b]$

Proof   First, for simplicity of notation, we shall write

$$L(x) = \prod_{k=0}^{N} (x - x_k) = (x - x_0)(x - x_1) \cdots (x - x_N)$$

Now, for $x \in [a, b]$, but not one of the nodes, consider the function E defined by

$$E(t) = f(t) - p(t) - cL(t)$$

where the *constant* c is chosen so that $E(x) = 0$. It now follows that $E(t)$ vanishes at the $N + 2$ distinct points $x_0, x_1, \ldots, x_N$, and x. By Rolle's theorem, between each successive pair of these there is a point at which E' vanishes. Repeating this argument for $E', E'', \ldots, E^{(N)}$ we eventually deduce that there is a point $\xi \in [a, b]$ such that

$$E^{(N+1)}(\xi) = 0$$

From the definition of E, this yields

$$f^{(N+1)}(\xi) - p^{(N+1)}(\xi) - cL^{(N+1)}(\xi) = 0$$

But, p is a polynomial of degree at most N, and so $p^{(N+1)} \equiv 0$, and $L^{(N+1)} \equiv (N + 1)!$ It follows that

$$c = f^{(N+1)}(\xi)/(N + 1)!$$

which, combined with the fact that $E(x) = 0$, completes the proof. ■

It should be pointed out here that Theorem 2 remains valid for x outside $[a, b]$ with just a slight adjustment to the range of possible values of ξ. In such a case, this theorem would cease being about *interpolation*. The process of *extrapolation*, in which we attempt to obtain values of a function outside the spread of the data, is numerically much less satisfactory since the error increases rapidly as x moves away from the interval $[a, b]$.

To see this, it is sufficient to study a graph of the function $L(x)$. In Figure 4.1, we have a graph of this function for the nodes $1, 1.5, 2, 2.5, 3$ so that

$$L(x) = (x - 1)(x - 1.5)(x - 2)(x - 2.5)(x - 3)$$

We see that the magnitude of this function remains small for $x \in [1, 3]$, but that it grows rapidly outside this interval. If $f^{(N+1)}$ varies only slowly then this is an accurate reflection of the behavior of the error term for Lagrange interpolation using these same nodes.

Generally speaking, we would expect the accuracy to improve as the number of nodes increases. Theorem 2 bears this out to some extent since the $(N + 1)!$ is likely to dominate the numerator in (4.7) *provided that* the higher-order derivatives of f remain bounded. We shall see later that this need not be the case in practice.

Unfortunately, it is also true that the inclusion of additional data points in the Lagrange interpolation formula is not a simple matter.

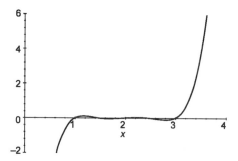

Figure 4.1   Growth of the Lagrange error

Example 3   Repeat the computation of Example 2 with the additional data point $f(1.4) = 0.3365$

The calculation of Example 2 is of no further benefit here and we must start afresh. Taking the nodes in numerical order, we get

$$p(1.2) = 0.3365\frac{(0.2)(-0.3)(-0.8)}{(0.4)(-0.1)(-0.6)} + 0.4055\frac{(0.2)(-0.2)(-0.8)}{(0.5)(0.1)(-0.5)}$$

$$+ 0.6931\frac{(0.2)(-0.2)(-0.3)}{(1)(0.6)(0.5)}$$

$$= 0.1817$$

to four decimal places.

The source of the data here was the function $\ln x$, and the true value $\ln 1.2 \approx 0.1823$ so that the actual error is around 6×10^{-4}. From (4.7), we can obtain a bound for this error as follows.

$$\ln 1.2 - p(1.2) = \frac{(0.2)(-0.2)(-0.3)(-0.8)}{4!}f^{(4)}(\xi)$$

for some ξ between 1 and 2. In this case, $f^{(4)}(x) = -3!/x^4$ and for $x \in [1,2]$, we deduce that $\left|f^{(4)}(x)\right| \leq 6$. It follows that the error is bounded by $\frac{0.0096}{24}(6) = 0.0024$.

The true error is *much* smaller than the predicted error. Of course this is likely since we computed the error *bound* using the worst-case scenario that $\xi = 1$.

There are many different ways of representing this interpolation polynomial, and we shall see some others in the next section. The principal motivation is to find convenient forms which can be implemented efficiently and are easily programmed. The Lagrange form does not fulfil either of these criteria. Since the polynomial itself is unique, these alternatives are just different ways of writing down the same thing. Therefore we do not need to repeat the error analysis.

The most important aspects of the Lagrange interpolation polynomial are therefore that it provides a relatively easy way to establish the existence and uniqueness of the polynomial, and it allows a simpler derivation of the error formula.

As well as being able to choose the representation of the interpolation polynomial, we may be in a position to choose the nodes themselves. Consideration of the remainder term (4.7) shows that it could be advantageous to choose the interpolation points so that $|L(x)|$ is kept as small as possible over the whole interval of interest. It is precisely this objective which leads to the choice of the Chebyshev interpolation points. For details of Chebyshev interpolation, the reader is referred to a more advanced numerical analysis text. Several such are listed among the references and further reading.

Exercises: Section 4.2

1 Find the Lagrange interpolation polynomial for the data

x	1	2	4
$f(x)$	3	2	1

Use this to estimate $f(1.5)$.

2 Repeat Exercise 1 with the additional data $f(0) = 4, f(3) = 1$.

3 Show that the Vandermonde determinant

$$V = \begin{vmatrix} 1 & x_0 & x_0^2 & \cdots & x_0^N \\ 1 & x_1 & x_1^2 & \cdots & x_1^N \\ & \cdots & & & \cdots \\ 1 & x_N & x_N^2 & \cdots & x_N^N \end{vmatrix}$$

does not vanish for distinct nodes $x_0, x_1, \ldots, x_N$. (**Hint:** Show that $(x_j - x_k)$ is a factor of V whenever $j \neq k$.)

4 Consider the following table of values of the cosine function:

x	0.0	0.1	0.2	0.3	0.4	0.5	0.6	0.7	0.8
$\cos x$	1.0000	0.9950	0.9801	0.9553	0.9211	0.8776	0.8253	0.7648	0.6967

Write down the Lagrange interpolation polynomial using the nodes $0.0, 0.1$ and 0.2, and then using 0.3 as well. Estimate the value of $\cos 0.14$ using each of these polynomials.

5 Obtain error bounds for the approximations in Exercise 4.

6 Show that for any $x \in (0.0, 0.8)$, the Lagrange interpolation quadratic using the 3 nearest nodes from the table in Exercise 4 results in an error less than 6.5×10^{-5}.

7 Show that if $x \in (0.1, 0.7)$ then the Lagrange interpolation polynomial using the 4 closest points in the table in Exercise 4 results in an error less than 5×10^{-6}.

8 Repeat Exercise 5 for the function $\ln(1 + x)$ for the same nodes. How many points will be needed to ensure that the Lagrange interpolation polynomial will introduce no new errors to 4 decimal places?

4.3 Difference representations

In this section, we are concerned with alternative representations of interpolation polynomials which may be more convenient and efficient to program, and to use in practice. All the representations we consider can be derived from the use of *divided differences*. We begin with Newton's divided difference representation. Subsequently, we shall look briefly at some *finite difference* interpolation formulas. These are special cases of the divided difference formula for the situation where all the nodes are equally spaced.

4.3.1 Divided difference interpolation

Before discussing their use in interpolation, we must define divided differences.

Definition 1 **The zero-th order divided difference of f at x_k is defined by**

$$f[x_k] = f(x_k)$$

First-order divided differences at 2 points x_j, x_k are then defined by

$$f[x_j, x_k] = \frac{f[x_k] - f[x_j]}{x_k - x_j} = f[x_k, x_j] \qquad (4.8)$$

Higher-order divided differences are defined recursively by

$$f[x_k, x_{k+1}, \ldots, x_{k+m}] = \frac{f[x_{k+1}, \ldots, x_{k+m}] - f[x_k, \ldots, x_{k+m-1}]}{x_{k+m} - x_k} \qquad (4.9)$$

We note that (4.8) implies a connection between first divided differences and first derivatives. Indeed in developing the secant iteration in Chapter 2, we used the approximation

$$f'(x_n) \approx \frac{f(x_n) - f(x_{n-1})}{x_n - x_{n-1}}$$

which is to say

$$f'(x_n) \approx f[x_{n-1}, x_n]$$

Further, by the mean value theorem, it follows that

$$f[x_{n-1}, x_n] = \frac{f(x_n) - f(x_{n-1})}{x_n - x_{n-1}} = f'(\xi)$$

for some point ξ between x_{n-1} and x_n.

We shall see shortly that there is a more general connection between divided differences and derivatives of the same order.

Example 4 **Compute the differences** $f[x_0, x_1]$, $f[x_0, x_2]$, $f[x_0, x_1, x_2]$, **and** $f[x_1, x_0, x_2]$ **for the data**

k	0	1	2
x_k	1	2	4
$f(x_k)$	3	2	1

Applying (4.8),

$$f[x_0, x_1] = \frac{f[x_1] - f[x_0]}{x_1 - x_0} = \frac{2 - 3}{2 - 1} = -1$$

$$f[x_0, x_2] = \frac{f[x_2] - f[x_0]}{x_2 - x_0} = \frac{1 - 3}{4 - 1} = -\frac{2}{3}$$

Now using (4.9)

$$f[x_0, x_1, x_2] = \frac{f[x_1, x_2] - f[x_0, x_1]}{x_2 - x_0} = \frac{-1/2 - (-1)}{3} = \frac{1}{6}$$

$$f[x_1, x_0, x_2] = \frac{f[x_0, x_2] - f[x_1, x_0]}{x_2 - x_1} = \frac{-2/3 - (-1)}{2} = \frac{1}{6}$$

We note that in Example 4, $f[x_0, x_1, x_2] = f[x_1, x_0, x_2]$. This *independence of order* is a general property of divided differences. There are several ways to establish this result. One of the algebraically least complicated appeals to the uniqueness of the interpolation polynomial. Direct proofs tend to be conceptually simpler but algebraically more intricate. For details of any of these proofs, the reader should consult a more advanced numerical analysis text. The first approach will be described once we have developed the basics of divided difference interpolation.

From the definition of first-order divided differences (4.8), we see that

$$f(x) = f[x_0] + (x - x_0) f[x_0, x] \tag{4.10}$$

Substituting (4.9) into (4.10) with $k = 0$ and using increasing values of m, we then get

$$
\begin{aligned}
f(x) &= f[x_0] + (x - x_0)f[x_0, x] \\
&= f[x_0] + (x - x_0)f[x_0, x_1] + (x - x_0)(x - x_1)f[x_0, x_1, x] \\
&= \cdots \\
&= f[x_0] + (x - x_0)f[x_0, x_1] + (x - x_0)(x - x_1)f[x_0, x_1, x_2] \\
&\quad + \cdots + (x - x_0)\cdots(x - x_{N-1})f[x_0, x_1, \ldots, x_N] \\
&\quad + (x - x_0)\cdots(x - x_N)f[x_0, x_1, \ldots, x_N, x]
\end{aligned}
\tag{4.11}
$$

If we now let p be the *polynomial* consisting of all but the last term of this expression:

$$
\begin{aligned}
p(x) &= f[x_0] + (x - x_0)f[x_0, x_1] + (x - x_0)(x - x_1)f[x_0, x_1, x_2] \\
&\quad + \cdots + (x - x_0)\cdots(x - x_{N-1})f[x_0, x_1, \ldots, x_N]
\end{aligned}
\tag{4.12}
$$

then p is a polynomial of degree at most N, and if $0 \le k \le N$, we obtain

$$
\begin{aligned}
p(x_k) &= f[x_0] + (x_k - x_0)f[x_0, x_1] + (x_k - x_0)(x_k - x_1)f[x_0, x_1, x_2] \\
&\quad + \cdots + (x_k - x_0)\cdots(x_k - x_{k-1})f[x_0, x_1, \ldots, x_k]
\end{aligned}
\tag{4.13}
$$

since all higher-degree terms have the factor $(x_k - x_k)$.

This enables us to prove the following theorem.

Theorem 3   The polynomial given by (4.12) satisfies the interpolation conditions

$$
p(x_k) = f(x_k) \qquad (k = 0, 1, \ldots, N)
$$

Proof   We shall proceed by induction. First (4.13), with $k = 0$ reduces to just

$$
p(x_0) = f[x_0] = f(x_0)
$$

as desired. Next suppose the result holds for $k = 0, 1, \ldots, m$ for some $m < N$. Using (4.11) with $N = m$ and $x = x_{m+1}$ we get

$$
\begin{aligned}
f(x_{m+1}) &= f[x_0] + (x_{m+1} - x_0)f[x_0, x_1] + (x_{m+1} - x_0)(x_{m+1} - x_1)f[x_0, x_1, x_2] \\
&\quad + \cdots + (x_{m+1} - x_0)\cdots(x_{m+1} - x_{m-1})f[x_0, x_1, \ldots, x_m] \\
&\quad + (x_{m+1} - x_0)\cdots(x_{m+1} - x_m)f[x_0, x_1, \ldots, x_m, x_{m+1}]
\end{aligned}
$$

which, using (4.13), is $p(x_{m+1})$. This completes the induction step and, therefore, the proof. ∎

It follows, by the uniqueness of the Lagrange interpolation polynomial, that this formula (4.12) is a rearrangement of the Lagrange polynomial. This particular form is known as *Newton's divided difference interpolation polynomial*. It is a particularly useful form of the polynomial as it allows the data points to be introduced one at a time without any of the waste of effort this entails for the Lagrange formula.

Another consequence of (4.12) is that the divided differences $f[x_0, x_1, \ldots, x_k]$ each represent the leading (degree k) coefficient of the interpolation polynomial which agrees with f at the nodes $x_0, x_1, \ldots, x_k$. By the uniqueness of this polynomial, this coefficient must be independent of the order of the nodes. It is this observation which establishes that divided differences depend only on the *set* of nodes, *not on their order.*

As yet another consequence of the uniqueness of the interpolation polynomial, we can deduce an important relation connecting divided differences and derivatives. We have already observed that first divided differences are approximations to first derivatives, or, using the Mean Value Theorem, *are* first derivatives at some 'mean value' point.

Subtracting equation (4.12) from (4.11), we obtain

$$f(x) - p(x) = (x - x_0) \cdots (x - x_N) f[x_0, x_1, \ldots, x_N, x]$$

but from Theorem 2, and, specifically, (4.7), we already know that

$$f(x) - p(x) = \frac{(x - x_0)(x - x_1) \cdots (x - x_N)}{(N + 1)!} f^{(N+1)}(\xi)$$

It follows that

$$f[x_0, x_1, \ldots, x_N, x] - \frac{f^{(N+1)}(\xi)}{(N + 1)!}$$

where the 'mean value point' ξ lies somewhere in the interval spanned by $x_0, x_1, \ldots, x_N$, and x.

Example 5 Use Newton's divided difference formula to estimate $\ln 1.2$ and $\ln 1.7$ from the following data

x	1.0	1.4	1.5	2.0
$\ln x$	0.0000	0.3365	0.4055	0.6931

Now, for $x = 1.2$ taking the data in the order given, we obtain the following table of divided differences:

k	x_k	$f[x_k]$	$f[x_k, x_{k+1}]$	$f[x_k, x_{k+1}, x_{k+2}]$	$f[x_k, x_{k+1}, x_{k+2}, x_{k+3}]$
0	1.0	0.0000	0.8413	−0.3026	0.1113
1	1.4	0.3365	0.6900	−0.1913	
2	1.5	0.4055	0.5752		
3	2.0	0.6931			

Here, for example, the entry $f[x_1,x_2] = 0.6900$ results from $(0.4055 - 0.3365)/(1.5 - 1.4)$. From the table of differences we obtain the successive approximations to 4 decimal places

$$f[x_0] + (x - x_0)f[x_0,x_1] = 0.0000 + (1.2 - 1.0)(0.8413)$$
$$= 0.16826 \approx 0.1683$$

and then,

$$0.16826 + (x - x_0)(x - x_1)f[x_0,x_1,x_2]$$
$$+ (1.2 - 1.0)(1.2 - 1.4)(-0.3026)$$
$$= 0.18036 \approx 0.1804$$

and, finally,

$$0.18036 + (1.2 - 1.0)(1.2 - 1.4)(1.2 - 1.5)(0.1113) = 0.1817$$

One immediately apparent aspect of this process is the ease of introducing new data points into the calculation, and the possibility of proceeding in an iterative manner until successive approximations to the required value agree to within some specified accuracy. This provides a much more practical way of performing polynomial interpolation which avoids the need for finding an error bound – often a difficult task in practice.

For $x = 1.7$, it is desirable to change the order in which the nodes are used. Taking $x_0 = 1.5$ which is the nearest one to 1.7, we could then take either 1.4 or 2.0 next. Using $x_1 = 2.0$ means that we are using interpolation at this stage since $1.7 \in (1.5, 2.0)$. The resulting table of differences is then:

k	x_k	$f[x_k]$	$f[x_k,x_{k+1}]$	$f[x_k,x_{k+1},x_{k+2}]$	$f[x_k,x_{k+1},x_{k+2},x_{k+3}]$
0	1.5	0.4055	0.5752	-0.1913	0.1114
1	2.0	0.6931	0.5943	-0.2470	
2	1.4	0.3365	0.8413		
3	1.0	0.0000			

The successive approximations thus obtained are

$$0.5205, \quad 0.5320, \quad 0.5300$$

The true value is $\ln 1.7 = 0.5306$ to 4 decimal places.

The following m-file for divided difference interpolation reorders the data for each point at which we wish to estimate the function. MATLAB's sort function is particularly helpful here.

Program	MATLAB m-file for divided difference interpolation using nearest nodes first:

```
function y=ddiff1(xdat,ydat,x)
% Computes Newton's divided difference formula
% For each element of x, uses nearest data points first
N=length(x);
M=length(xdat);
for k=1:N
% First sort the data arrays
    xtst=x(k);
    [xb,ind]=sort(abs(xtst-xdat));
    xsort=xdat(ind);
    D(:,1)=ydat(ind)';
% Compute the differences and approximation
  for j=1:M
    for i=1:M-j
       D(i,j+1)=(D(i+1,j)-D(i,j))/(xsort(i+j)-xsort(i));
    end
end % Completes computation of divided differences
% Next the interpolation computation
    xdiff=xtst-xsort;
    prod=1; prod holds current product of xdiff
    y(k)=0;
for i=1:M
    y(k)=y(k)+prod*D(1,i);
    prod=prod*xdiff(i);
  end
```

For the data of Example 5, the following MATLAB commands yield the results shown:

```
» X=[1 1.4 1.5 2.0];
» Y=log(X);
» x=[1.2,1.7];
» y=ddiff1(X,Y,x)
y =
  0.1817   0.5300
```

as we obtained earlier.

In the situation where we have many data points it may be desirable to limit the degree of the interpolation polynomial. This is easily incorporated into the code by restricting the variable M to be bounded by some maximum degree.

If we wish to evaluate Newton's formula at many points (in order to graph the function, for example) it might be more economical to avoid re-sorting the data for each argument. If the order of the data is not to be changed, then it is typically the case that *all* data points are used throughout. However if the number of data points is also large, the resulting graph may exhibit the natural tendency of a polynomial to 'wiggle'.

Example 6   **Graph the divided difference polynomial for the data**

x	1.0000	1.1000	1.3000	1.4000	1.5500	1.7500	1.8000	2.0000
y	0.0000	0.9530	2.6240	3.3650	4.3830	5.5960	5.8780	6.9310

over the interval $[1, 2]$

With data vectors X,Y and x=1:0.01:2, we can use the MATLAB commands

```
» y=ddiff1(X,Y,x);
» plot(X,Y,'k*',x,y,'k')
» axis([0.9 2.1 -0.5 7.5])
```

These generate the graph shown in Figure 4.2.

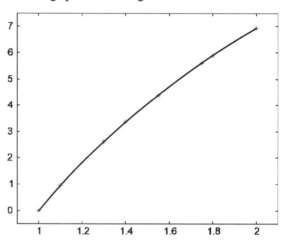

Figure 4.2 Divided difference interpolation

In this case, we see a smooth curve passing through all the data points. Polynomial interpolation appears highly satisfactory for this example.

We commented above that for graphing purposes it may be desirable to avoid re-sorting the data. The code used previously is easily modified for this purpose. (See the Exercises.) The computation for Example 7 used such an m-file.

Example 7 **Graph the interpolation polynomial for data from the function**

$$y = \frac{1}{1 + x^2}$$

using nodes at the integers in $[-5, 5]$

The MATLAB commands

```
» X=-5:5; Y=runge(X);
» x=-5:0.05:5;
» y=ddint(X,Y,x);
» plot(X,Y,'k*',x,y,'k')
» hold on;
» plot(x,runge(x))
```

where the m-file runge.m implements the given function, produce the graph shown in Figure 4.3.

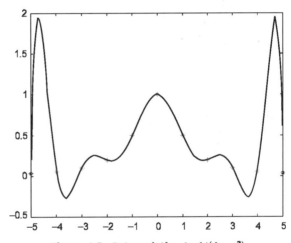

Figure 4.3 **Interpolation to** $1/(1 + x^2)$

We see that the interpolation polynomial and the 'true' function become quite distinct towards the ends of the range. The curves have reasonable resemblance to each other in the middle of the interval but as $|x|$ grows the polynomial behaviour of the interpolation formula cannot be controlled.

This is one case of a famous example due to Runge. If the number of (equally spaced) nodes is increased the oscillations of the polynomial become steadily wilder. Indeed, Runge showed that, for this example, the interpolation polynomials of increasing degree *diverge* for $|x| > 3.5$.

If the degree of the interpolating polynomial is restricted so that the nearest 4 nodes (to the current point) are used to generate a local cubic interpolating polynomial, then the resulting curve fits the original function much better. The commands

```
» plot(X,Y,'k*',x,runge(x))
» y1=interp1(X,Y,x,'cubic');
» hold on; plot(x,y1,'k')
```

produce the graph below. It is difficult to see much difference between the two curves.

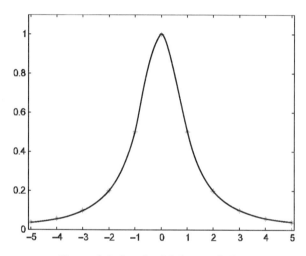

Figure 4.4   Local cubic interpolation

The function interp1 is a built-in MATLAB interpolation function of which more will be said later.

In Example 7, the data points were equally spaced. In such circumstances, the divided difference formula can be rewritten in special forms.

There are efficient ways of implementing divided difference interpolation to reduce the computational complexity. Among these, perhaps the most widely used is the algorithm of Aitken, which computes the values of the interpolation polynomials directly without the need for *explicit* computation of the divided differences. The details of Aitken's algorithm are left to subsequent, more advanced courses. We turn our attention to the special case of finite difference interpolation for data at equally spaced nodes.

4.3.2 Finite difference interpolation

In this section we shall assume that the interpolation points are spaced at regular intervals of length h, so that

$$x_k = x_0 + kh, \qquad k = 0, \pm 1, \pm 2, \ldots \qquad (4.14)$$

There are many different finite difference interpolation formulas designed to fit slightly different situations. We shall consider just two of these in order to give a flavour of the subject. For simplicity we shall describe both of these formulas in terms of *forward differences*. There are also central and backward differences, but we shall restrict our attention to the forward differences.

Definition 2   We define *forward differences* of f at x_k as follows:

$$
\begin{aligned}
\textbf{\textit{First difference:}} \quad & \Delta f(x_k) = f(x_{k+1}) - f(x_k) \\
\textbf{\textit{Second difference:}} \quad & \Delta^2 f(x_k) = \Delta f(x_{k+1}) - \Delta f(x_k) \\
& = f(x_{k+2}) - 2f(x_{k+1}) + f(x_k) \\
\textbf{\textit{n-th difference}} \quad & \Delta^n f(x_k) = \Delta^{n-1} f(x_{k+1}) - \Delta^{n-1} f(x_k)
\end{aligned}
$$

For example, the third difference $\Delta^3 f(x_0)$ is given by

$$
\begin{aligned}
\Delta^3 f(x_0) &= \Delta^2 f(x_1) - \Delta^2 f(x_0) \\
&= [f(x_3) - 2f(x_2) + f(x_1)] - [f(x_2) - 2f(x_1) + f(x_0)] \\
&= f(x_3) - 3f(x_2) + 3f(x_1) - f(x_0)
\end{aligned}
$$

Example 8   Compute $\Delta f(x_0), \Delta f(x_1), \Delta^2 f(x_0), \Delta^3 f(x_0)$ for the data

k	0	1	2	3
x_k	-0.5	0.0	0.5	1.0
$f(x_k)$	2	3	5	8

Here $\Delta f(x_0) = f(x_1) - f(x_0) = 3 - 2 = 1$, and $\Delta f(x_1) = f(x_2) - f(x_1) = 5 - 3 = 2$.

Then, $\Delta^2 f(x_0) = \Delta f(x_1) - \Delta f(x_0) = 2 - 1 = 1$.

Also $\Delta^3 f(x_0) = f(x_3) - 3f(x_2) + 3f(x_1) - f(x_0) = 8 - 3(5) + 3(3) - 2 = 0$.

Typically, it would be easier to compute the difference table so that each new difference would just be a difference of two differences of lower order, as in the definition. For this example, the difference table would be:

x	f	Δf	$\Delta^2 f$	$\Delta^3 f$
-0.5	2			
		1		
0.0	3		1	
		2		0
0.5	5		1	
		3		
1.0	8			

where each difference entry is just the difference of its two neighbours to the left.

It is not too difficult to show by induction that

$$\Delta^n f(x_k) = \sum_{j=0}^{n} (-1)^j \binom{n}{j} f\left(x_{k+n-j}\right) \tag{4.15}$$

Also, for equally spaced nodes, we can relate divided and forward differences by

$$f[x_k, x_{k+1}, \ldots, x_{k+n}] = \frac{\Delta^n f(x_k)}{h^n n!} \tag{4.16}$$

Now, writing $x = x_0 + sh$, $f_k = f(x_k)$ and $\Delta^n f_k = \Delta^n f(x_k)$, and using (4.16) to substitute into the divided difference interpolation formula (4.12), we get

$$f(x_0 + sh) \approx f[x_0] + (x - x_0) f[x_0, x_1] + (x - x_0)(x - x_1) f[x_0, x_1, x_2] + \cdots$$

$$= f_0 + sh \frac{\Delta f_0}{h} + sh(s-1)h \frac{\Delta^2 f_0}{2! h^2} + \cdots \tag{4.17}$$

$$= f_0 + s\Delta f_0 + \frac{s(s-1)}{2!} \Delta^2 f_0 + \cdots$$

Equation (4.17) is *Newton's forward difference interpolation formula.*

Example 9   Use Newton's forward difference interpolation formula to estimate ln 1.13 from the following table of data

x	1.0	1.1	1.2	1.3	1.4	1.5	1.6	1.7
$\ln x$	0.0000	0.0953	0.1823	0.2624	0.3365	0.4055	0.4700	0.5306

In much the same way as for divided differences, we begin by forming the difference table:

x	f	Δf	$\Delta^2 f$	$\Delta^3 f$	$\Delta^4 f$
1.0	0.0000				
		0.0953			
1.1	0.0953		−0.0083		
		0.0870		0.0014	
1.2	0.1823		−0.0069		−0.0005
		0.0801		0.0009	
1.3	0.2624		−0.0060		0.0000
		0.0741		0.0009	
1.4	0.3365		−0.0051		−0.0003
		0.0690		0.0006	
1.5	0.4055		−0.0045		0.0000
		0.0645		0.0006	
1.6	0.4700		−0.0039		
		0.0606			
1.7	0.5306				

The forward difference formula (4.17) requires that we choose a base-point x_0 and then introduce further interpolation points $x_1, x_2, \ldots$ in turn. We are not in a position to use nodes in the order of their proximity to the point of interest x, in this case 1.13.

First, consider the choice $x_0 = 1.1$ which is the closest node to $x = 1.13$. With $h = 0.1$, we have $x = x_0 + 0.3h$ so that we take $s = 0.3$ in (4.17). The successive estimates as we incorporate more terms of the interpolation formula are therefore:

$$\ln 1.13 \approx f_0 = 0.0953$$
$$\ln 1.13 \approx f_0 + s\Delta f_0 = 0.0953 + (0.3)(0.0870) = 0.121\,4$$

$$\ln 1.13 \approx f_0 + s\Delta f_0 + \frac{s(s-1)}{2}\Delta^2 f_0$$
$$= 0.121\,4 + \frac{(0.3)(-0.7)}{2}(-0.0069)$$
$$= 0.122125$$

and then

$$\ln 1.13 \approx f_0 + s\Delta f_0 + \frac{s(s-1)}{2}\Delta^2 f_0 + \frac{s(s-1)(s-2)}{3!}\Delta^3 f_0$$
$$= 0.122125 + \frac{(0.3)(-0.7)(-1.7)}{6}(0.0009)$$
$$= 0.122178$$

and since, to the accuracy of the data, the next difference is zero, no improvement will be obtained by including further terms. Higher-order differences could be computed and incorporated into the approximation but, in this case, further differences would be dominated by the effect of the roundoff errors in the original data. (In this case, for example, the next term – corresponding to the fifth difference – would be negative even though the true value, $\ln 1.13 = 0.12221763$, is greater than the current estimate.)

We have therefore achieved as much accuracy as this data allows with $x_0 = 1.1$.

If, instead, we take $x_0 = 1.0$ with $s = 1.3$, then the first 2 estimates will be less accurate since they would be based on *extrapolation.* The approximate values obtained are

$$\ln 1.13 \approx f_0 = 0.0000$$
$$\ln 1.13 \approx f_0 + s\Delta f_0 = 0.0000 + (1.3)(0.0953) = 0.12389$$
$$\ln 1.13 \approx f_0 + s\Delta f_0 + \frac{s(s-1)}{2}\Delta^2 f_0$$
$$= 0.12389 + \frac{(1.3)(0.3)}{2}(-0.0083)$$
$$= 0.1222715$$

and then

$$\ln 1.13 \approx f_0 + s\Delta f_0 + \frac{s(s-1)}{2}\Delta^2 f_0 + \frac{s(s-1)(s-2)}{3!}\Delta^3 f_0$$
$$= 0.1222715 + \frac{(1.3)(0.3)(-0.7)}{6}(0.0014)$$
$$= 0.1222078$$

which has a final error of about 10^{-5}.

The improved estimate here is due to the fact that this final value is obtained using the nodes 1.0, 1.1, 1.2, and 1.3 which are the four closest points among the data.

At the beginning of the computation, we do not know how many terms, and therefore which nodes, are to be used. It is desirable to find a way of combining the advantages of equal spacing of the nodes with those of using the closest points first.

We thus seek a *finite difference* formula such that $x \in (x_0, x_1)$, and the interpolation points are used in the order $x_0, x_1, x_{-1}, x_2, x_{-2}, \ldots$ Like the forward difference formula, this can be achieved starting with the divided difference formula for these same points:

$$f(x_0 + sh) \approx f[x_0] + (x - x_0)f[x_0, x_1] + (x - x_0)(x - x_1)f[x_{-1}, x_0, x_1] + \cdots$$
$$= f_0 + s\Delta f_0 + \frac{s(s-1)}{2!}\Delta^2 f_{-1} + \frac{s(s-1)(s+1)}{3!}\Delta^3 f_{-1} \qquad (4.18)$$

The general term of this expansion for an *even*-order difference is

$$\frac{(s+m-1)(s+m-2)\cdots(s+1)s(s-1)\cdots(s-m)}{(2m)!}\Delta^{2m}f_{-m}$$

while that for an *odd*-order difference is

$$\frac{(s+m)(s+m-1)\cdots(s+1)s(s-1)\cdots(s-m)}{(2m+1)!}\Delta^{2m+1}f_{-m}$$

The effect is that we use entries in the difference table which lie closest to the horizontal line corresponding to the required value x.

This formula (4.18) is called *Gauss' central difference formula*. It is often written using central differences such as $\delta f_{1/2} = f_1 - f_0$ rather than the forward differences we have used. For our purposes the forward difference version will suffice.

Example 10   Use the data of Example 9 to estimate $\ln 1.34$ using the central difference formula

The difference table is as before:

x	f	Δf	$\Delta^2 f$	$\Delta^3 f$	$\Delta^4 f$
1.0	0.0000				
		0.0953			
1.1	0.0953		−0.0083		
		0.0870		0.0014	
1.2	0.1823		−0.0069		−0.0005
		0.0801		0.0009	
1.3	**0.2624**		**−0.0060**		0.0000
		0.0741		**0.0009**	
1.4	0.3365		−0.0051		−0.0003
		0.0690		0.0006	
1.5	0.4055		−0.0045		0.0000
		0.0645		0.0006	
1.6	0.4700		−0.0039		
		0.0606			
1.7	0.5306				

With $x_0 = 1.3$, we use the entries in **bold** type, which corresponds to taking the nodes in the order 1.3, 1.4, 1.2, 1.5. The successive approximations obtained are then

$$\ln 1.34 \approx f_0 = 0.2624$$
$$\ln 1.34 \approx f_0 + s\Delta f_0 = 0.2624 + (0.4)(0.0741) = 0.29204$$
$$\ln 1.34 \approx f_0 + s\Delta f_0 + \frac{s(s-1)}{2}\Delta^2 f_{-1}$$
$$= 0.29204 + \frac{(0.4)(-0.6)}{2}(-0.0060) = 0.29276$$

and

$$\ln 1.34 \approx f_0 + s\Delta f_0 + \frac{s(s-1)}{2}\Delta^2 f_{-1} + \frac{s(s-1)(s+1)}{3!}\Delta^3 f_{-1}$$

$$= 0.29276 + \frac{(1.4)(0.4)(-0.6)}{6}(0.0009)$$

$$= 0.292710$$

The true value is $\ln 1.34 = 0.292670$ so that our approximation is accurate to the 4 decimal places of the data, which is as much accuracy as we have any right to expect.

In introducing this approach to finite difference interpolation, we assumed that x_0 was chosen so that $x \in (x_0, x_1)$. If x lies much closer to x_1 than to x_0, then the order of the nodes is not exactly as we desired. In such a situation, it would be preferable to choose x_0 so that $x \in (x_{-1}, x_0)$ and then use the data points in the order x_0, x_{-1}, x_1, x_{-2}, ... This yields a similar formula to (4.18), known as Gauss' *backward* central difference formula. (In this sense, (4.18) is known as the *forward* central difference formula.)

We do not pursue any of these further here. We have already seen that there are practical difficulties with polynomial interpolation. We saw in the previous section that locally restricting the degree of the polynomial used can result in much better approximation to a smooth function – but such an approximation does not reflect the smoothness of the function being approximated.

The next section addresses this difficulty through the use of splines.

Exercises:
Section 4.3

1 Compute the divided differences $f[x_0, x_1], f[x_0, x_2]$, and $f[x_0, x_1, x_2]$ for the data:

k	0	1	2
x_k	0	1	3
$f(x_k)$	1	2	3

2 Write down the quadratic polynomial p which has the values $p(0) = 1$, $p(1) = 2$, $p(3) = 3$. (**Hint**: use Exercise 1.)

3 Use the divided difference interpolation formula to estimate $f(0.24)$ from the data

$$f(0) = 6.021, \quad f(0.2) = 6.232, \quad f(0.3) = 6.335$$
$$f(0.4) = 6.435, \quad f(0.6) = 6.628$$

4 Use divided difference interpolation to obtain values of $\cos 0.14, \cos 0.35$ and $\cos 0.68$ from the following data:

x	0.0	0.1	0.2	0.3	0.4
$\cos x$	1.0000	0.9950	0.9801	0.9553	0.9211

x	0.5	0.6	0.7	0.8
$\cos x$	0.8776	0.8253	0.7648	0.6967

(You will probably find this easiest using MATLAB.)

5 Write a MATLAB m-file for divided difference interpolation which uses all the data, in the same order, at all points. Use this to plot the divided difference interpolation polynomial for the data in Exercise 4. Also plot the error function for this interpolation polynomial over the interval $[0, 1]$. (Note the behaviour in the *extrapolation* region.)

6 Compute the forward difference table for the data:

x	0	1	2	3
$f(x)$	1	0	1	2

7 Write down the interpolation polynomial for the data of Exercise 6. Estimate $f(1.3)$ using

(a) the forward difference interpolation formula, and
(b) the central difference interpolation formula

8 Use induction to prove (4.16), namely

$$f[x_k, x_{k+1}, \ldots, x_{k+n}] = \frac{\Delta^n f(x_k)}{h^k k!}$$

9 Show that

$$\Delta^n f(x_k) = \sum_{j=0}^{n} (-1)^j \binom{n}{j} f\left(x_{k+n-j}\right)$$

10 Use Newton's forward difference formula to estimate $\cos 0.14$, and the central difference formula to estimate $\cos 0.35$ and $\cos 0.68$ from the data in Exercise 4.

11 Write a MATLAB m-file to implement central difference interpolation. Test it by finding approximate values for $f(0.24)$ and $f(0.37)$ from the following data

$$f(0) = 6.021, \quad f(0.1) = 6.128, \quad f(0.2) = 6.232,$$
$$f(0.3) = 6.335, \quad f(0.4) = 6.435, \quad f(0.5) = 6.532, \quad f(0.6) = 6.628$$

4.4 Splines

We observed at the beginning of this chapter that interpolation using a single polynomial is not always appropriate for the approximate evaluation of a function. Using divided difference or finite difference formulas can alleviate some of the difficulty by using only those data points that are close to the current point of interest. Effectively such methods approximate the function by different low-degree polynomials in different intervals; that is, they use *piecewise polynomials*.

The problem with the piecewise polynomials generated by, say, divided difference interpolation using cubic pieces (which would mean using the 4 nearest nodes to each point) is that the transition from one piece to the next is typically not smooth. This situation is illustrated in Figure 4.5.

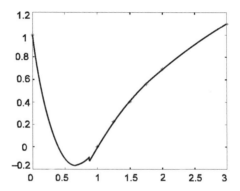

Figure 4.5 Piecewise cubic interpolation

The nodes used were 0, 1, 1.25, 1.5, 1.75, 2, 3. The data are plotted along with the divided difference piecewise cubic using the nearest four nodes. There is an abrupt change (indeed, a discontinuity) at the point at which the node 0 is dropped from the set being used. For these nodes, this happens as soon as $|x - 1.75| < |x|$, which occurs at $7/8$.

The idea behind *spline* interpolation is to eliminate such loss of smoothness at the transitions. We first need to know what splines are.

Definition 3   Let $a = x_0 < x_1 < \cdots < x_n = b$. A function $s : [a, b] \to \Re$ is a *spline* or *spline function* of degree m with *knots* (or *nodes*, or *interpolation points*) $x_0, x_1, \ldots, x_n$ if

1 s is a piecewise polynomial such that, on each subinterval $[x_k, x_{k+1}]$, s has degree at most m, and
2 s is $m - 1$ times differentiable everywhere.

Such a function is therefore defined by a different polynomial formula on each of the 'knot intervals' $[x_k, x_{k+1}]$. It can therefore be differentiated as often as we like at all points between the knots. The important extra condition is that its first $m - 1$ derivatives are also continuous at each knot.

It is important to note that for spline interpolation it is essential that the knots be in strictly increasing order. (Splines *can* be defined with coincident knots, but we shall not consider such cases here.) The name 'spline' comes from the old draftsman's tool of the same name. A spline was a thin flexible piece of wood which was used to enable the draftsman to construct a smooth curve through a set of specific points. A pin would be inserted into the drawing at each of these points, and the spline would then be tied to each of the pins – hence the term 'knots'. The spline would then assume a smooth curve, with minimal stored energy, passing through – or 'interpolating' – these points. In that particular case, the shape assumed by the physical spline turns out to be that of a *natural* cubic spline. We shall study this most important special case of spline interpolation shortly.

The basic idea of a spline is quite familiar, especially at the simplest level of *linear splines*. In graphical representation of common economic indicators such as the *Financial Times* or *Dow–Jones* indices, we see the familiar pictures of straight lines joining the dots that represent the actual data points. These straight lines represent the linear interpolation spline for that data. Thus the idea of spline interpolation could be viewed as the mathematical equivalent of childhood dot-to-dot drawings.

Example 11 **Show that the following function is a spline of degree 2, a** *quadratic spline*

$$s(x) = \begin{cases} x^2 + x & x \in [-1, 0] \\ x & x \in [0, 2] \\ x^2 - 3x + 4 & x \in [2, 5] \end{cases}$$

All that is needed is to show that s and s' are continuous at the internal knots $0, 2$. Now

$$\begin{aligned} \text{As } x \to 0- \quad & s(x) \to 0 \quad s'(x) = 2x + 1 \to 1 \\ \text{As } x \to 0+ \quad & s(x) \to 0 \quad s'(x) = 1 \end{aligned}$$

and so s, s' are continuous at $x = 0$. Similarly

$$\begin{aligned} \text{As } x \to 2- \quad & s(x) \to 2 \quad s'(x) = 1 \\ \text{As } x \to 2+ \quad & s(x) \to 2 \quad s'(x) = 2x - 3 \to 1 \end{aligned}$$

establishing the desired continuity at $x = 2$.

This particular spline is illustrated in Figure 4.6.

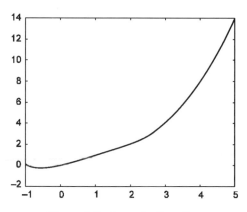

Figure 4.6 A quadratic spline

The continuity of the function and its slope are apparent from the graph. It is also apparent that the curvature of the function has two abrupt changes at the internal knots.

Example 12   Is the following function a spline?

$$s(x) = \begin{cases} x^3 + 2x & x \in [-1, 0] \\ 2x + 2x^2 & x \in [0, 1] \\ x^3 - x^2 + 5x - 1 & x \in [1, 3] \end{cases}$$

If this function is to be a spline it must be a cubic spline. Therefore we must check continuity of s, s' and s'' at $x = 0, 1$.

At both points, s is continuous. (Both relevant components give $s(0) = 0$, and at $x = 1$, both give $s(1) = 4$.)

Similarly, it is easy to check that s' is continuous with $s'(0) = 2$, $s'(1) = 6$.

However, for $x < 0$, $s''(x) = 6x \to 0$ as $x \to 0$, while, for $x \in (0, 1)$, $s''(x) = 4$. It follows that s'' is not continuous at $x = 0$. Therefore s is *not* a cubic spline. (Note that s'' is continuous at $x = 1$ in this case.)

It turns out that interpolation using low-degree splines can provide very good and economical methods for the approximation of a function. We shall begin with linear splines and then go on to the important special case of cubic spline interpolation.

Specifically, a linear spline, or spline of degree 1, is a continuous function whose graph consists of pieces which are all straight lines. Suppose then we are given the values of a function f at the knots

$$a = x_0 < x_1 < \cdots < x_n = b$$

and that we seek the linear spline s which satisfies

$$s(x_k) = f(x_k) \qquad (k = 0, 1, \ldots, n) \tag{4.19}$$

In the knot interval $[x_k, x_{k+1}]$, s must be a polynomial of degree 1 passing through the points $(x_k, f(x_k))$ and $(x_{k+1}, f(x_{k+1}))$. The equation of this line is

$$y = f_k + \frac{f_{k+1} - f_k}{x_{k+1} - x_k} (x - x_k) \tag{4.20}$$

where, as before, we have used f_k to denote $f(x_k)$.

Equation (4.20) for $k = 0, 1, \ldots, n - 1$ can be used to define the required spline function s. In the case of linear spline interpolation, we thus see that it is easy to write down formulas for the coefficients of the various components of the spline function. For more general splines this is not so straightforward.

It is common to use the notation s_k for the component of s which applies to the interval $[x_k, x_{k+1}]$. For the linear spline above we then have

$$s_k(x) = a_k + b_k(x - x_k)$$

and from (4.20), we see that

$$a_k = f_k$$

and

$$b_k = \frac{f_{k+1} - f_k}{x_{k+1} - x_k} = f[x_k, x_{k+1}]$$

Example 13   **Find the linear spline which takes the following values**

x	0.0	0.2	0.3	0.5
$f(x)$	0.00	0.18	0.26	0.41

Here, $x_0 = 0.0$, $x_1 = 0.2$, $x_2 = 0.3$, $x_3 = 0.5$,

$$a_0 = f_0 = 0.00$$
$$a_1 = f_1 = 0.18$$
$$a_2 = f_2 = 0.26$$

and

$$b_0 = f[x_0, x_1] = \frac{0.18 - 0.00}{0.2 - 0.0} = 0.90$$
$$b_1 = f[x_1, x_2] = \frac{0.26 - 0.18}{0.3 - 0.2} = 0.80$$
$$b_2 = f[x_2, x_3] = \frac{0.41 - 0.26}{0.5 - 0.3} = 0.75$$

Therefore

$$s(x) = \begin{cases} 0.00 + 0.9(x - 0.0) = 0.9x & 0.0 \le x \le 0.2 \\ 0.18 + 0.8(x - 0.2) = 0.8x + 0.02 & 0.2 \le x \le 0.3 \\ 0.26 + 0.75(x - 0.3) = 0.75x + 0.035 & 0.3 \le x \le 0.5 \end{cases}$$

Note that there is no ambiguity in this definition despite including interior knots in two intervals. The spline is continuous at these knots, which ensures that the two definitions must agree at these points.

Note also that the linear spline is not even defined outside the range of the data. Extrapolation using a linear spline would be especially suspect. In the particular case of Example 13 the data were taken from $\ln(1 + x)$ and, in contrast to the last remark, the graphs of this function and the linear spline would be virtually indistinguishable over the interval $[0, 0.5]$. They would diverge rapidly outside this interval, however.

Probably the most commonly used spline functions are *cubic splines*. Following the style of our linear splines, we can write the components of a cubic spline in the form

$$s_k(x) = a_k + b_k(x - x_k) + c_k(x - x_k)^2 + d_k(x - x_k)^3 \qquad (4.21)$$

Now s must satisfy the interpolation conditions

$$s_k(x_k) = f(x_k), \quad s_k(x_{k+1}) = f(x_{k+1}) \qquad (k = 0, 1, \ldots, n-1) \qquad (4.22)$$

which guarantees the continuity of s. Also we require the first two derivatives to be continuous. These conditions can be written as

$$s_k'(x_{k+1}) = s_{k+1}'(x_{k+1}) \qquad (k = 0, 1, \ldots, n-2) \qquad (4.23)$$
$$s_k''(x_{k+1}) = s_{k+1}''(x_{k+1}) \qquad (k = 0, 1, \ldots, n-2) \qquad (4.24)$$

These equations give us a total of $4n - 2$ equations for the $4n$ coefficients. These equations are linear in the coefficients and so we have 2 *degrees of freedom*. There are several ways of using these. We shall consider just one, the *natural cubic spline*. The specifics of this aspect will emerge shortly.

First, the interpolation conditions $s_k(x_k) = f(x_k)$ immediately yield (just as for the linear spline)

$$a_k = f_k \qquad (k = 0, 1, \ldots, n-1) \qquad (4.25)$$

Substituting this into the remaining interpolation conditions we get

$$b_k(x_{k+1} - x_k) + c_k(x_{k+1} - x_k)^2 + d_k(x_{k+1} - x_k)^3 = f_{k+1} - f_k \qquad (4.26)$$

It will be convenient to denote the steplengths by h_k, so that

$$h_k = x_{k+1} - x_k$$

Substituting this into (4.26) and dividing by h_k, we now have, for $k = 0, 1, \ldots,$ $n - 1$,

$$b_k + c_k h_k + d_k h_k^2 = \frac{f_{k+1} - f_k}{h_k} = f[x_k, x_{k+1}] = \delta_k \qquad (4.27)$$

say.

Next, substituting into (4.23) and (4.24) gives

$$b_k + 2c_k h_k + 3d_k h_k^2 = b_{k+1} \qquad (k = 0, 1, \ldots, n - 2) \qquad (4.28)$$

and

$$2c_k + 6d_k h_k = 2c_{k+1} \qquad (k = 0, 1, \ldots, n - 2)$$

This last equation yields

$$d_k = \frac{c_{k+1} - c_k}{3h_k} \qquad (4.29)$$

and substituting (4.29) into (4.27) yields

$$b_k = \delta_k - c_k h_k - \frac{(c_{k+1} - c_k)h_k}{3}$$
$$= \delta_k - \frac{h_k}{3}(c_{k+1} + 2c_k) \qquad (4.30)$$

It now follows that if we can determine the coefficients c_k, then (4.29) and (4.30) can be used to complete the definition of the spline components. Substituting for b_k, b_{k+1} and d_k in (4.28), we arrive at a linear system of equations for the coefficients c_k, as follows:

$$\delta_k - \frac{h_k}{3}(c_{k+1} + 2c_k) + 2c_k h_k + (c_{k+1} - c_k)h_k = \delta_{k+1} - \frac{h_{k+1}}{3}(c_{k+2} + 2c_{k+1})$$

Collecting terms, we obtain

$$h_k c_k + 2(h_k + h_{k+1})c_{k+1} + h_{k+1}c_{k+2} = 3(\delta_{k+1} - \delta_k) \qquad (4.31)$$

for $k = 0, 1, \ldots, n - 3$. This represents a *tridiagonal* system of linear equations for the unknown coefficients c_k. (A tridiagonal system of linear equations is one whose matrix of coefficients has all elements zero except for its diagonal and the immediate subdiagonal and superdiagonal. Therefore only 3 consecutive unknowns appear in any one of the equations.) Such systems are easily solved, as we shall see shortly.

The system (4.31), however, has only $n - 2$ equations for the n unknowns. This is where we utilize the 2 degrees of freedom discussed earlier.

Definition 4 The *natural cubic spline* is defined by imposing the additional conditions

$$s''(a) = s''(b) = 0$$

This allows the spline function to be continued with straight lines outside the interval $[a, b]$ while maintaining its smoothness. This, of course, mimics the behaviour of the physical spline beyond the extreme knots.

For the natural cubic spline, we therefore obtain

$$s''(a) = s_0''(x_0) = 2c_0 = 0$$

while

$$s''(b) = s_{n-1}''(x_n) = 2c_{n-1} + 6d_{n-1}h_{n-1} = 0$$

Introducing the spurious coefficient $c_n = 0$, this last equation becomes

$$d_{n-1} = \frac{c_n - c_{n-1}}{3h_{n-1}}$$

which is to say that (4.29) remains valid for $k = n - 1$. It also extends the validity of (4.31) to $k = n - 2$.

The full tridiagonal system is now

$$H \begin{bmatrix} c_1 \\ c_2 \\ \vdots \\ c_{n-1} \end{bmatrix} = 3 \begin{bmatrix} \delta_1 - \delta_0 \\ \delta_2 - \delta_1 \\ \vdots \\ \delta_{n-1} - \delta_{n-2} \end{bmatrix}$$

where H is the tridiagonal matrix

$$\begin{bmatrix} 2(h_0 + h_1) & h_1 & & & & \\ h_1 & 2(h_1 + h_2) & h_2 & & & \\ & h_2 & 2(h_2 + h_3) & h_3 & & \\ & & \ddots & \ddots & \ddots & \\ & & & h_{n-3} & 2(h_{n-3} + h_{n-2}) & h_{n-2} \\ & & & & h_{n-2} & 2(h_{n-2} + h_{n-1}) \end{bmatrix}$$

We now have $n - 2$ equations in the $n - 2$ remaining unknowns.

Before looking at the question of implementing cubic spline interpolation, we consider one small example. Even from this example, it becomes apparent that this technique is not well-suited to 'hand' calculation!

Example 14   **Find the natural cubic spline which fits the data**

x	25	36	49	64	81
$f(x)$	5	6	7	8	9

There are 5 knots, so $n = 4$ and we should obtain a 3×3 tridiagonal system of linear equations for the coefficients c_1, c_2, c_3.

First, $a_k = f_k$ for $k = 0, 1, 2, 3$ so that

$$a_0 = 5, \ a_1 = 6, \ a_2 = 7, \ a_3 = 8$$

The steplengths are $h_0 = 11$, $h_1 = 13$, $h_2 = 15$, $h_3 = 17$ and the divided differences are therefore

$$\delta_0 = 1/11, \ \delta_1 = 1/13, \ \delta_2 = 1/15, \ \delta_3 = 1/17$$

The tridiagonal system is then

$$2(11 + 13)c_1 + 13c_2 = 3\left(\frac{1}{13} - \frac{1}{11}\right) = -0.0420$$

$$13c_1 + 2(13 + 15)c_2 + 15c_3 = 3\left(\frac{1}{15} - \frac{1}{13}\right) = -0.0308$$

$$15c_2 + 2(15 + 17)c_3 = 3\left(\frac{1}{17} - \frac{1}{15}\right) = -0.0235$$

In matrix terms:

$$\begin{bmatrix} 48 & 13 & 0 \\ 13 & 56 & 15 \\ 0 & 15 & 64 \end{bmatrix} \begin{bmatrix} c_1 \\ c_2 \\ c_3 \end{bmatrix} = \begin{bmatrix} -0.0420 \\ -0.0308 \\ -0.0235 \end{bmatrix}$$

This system has the solution

$$c_1 = -0.798 \times 10^{-3}, \quad c_2 = -0.284 \times 10^{-3}, \quad c_3 = -0.301 \times 10^{-3}$$

Then using (4.29) with $c_0 = c_4 = 0$, we get

$$d_0 = \frac{c_1 - c_0}{3h_0} = \frac{-0.798 \times 10^{-3}}{3(11)} = -2.22 \times 10^{-5}$$

$$d_1 = \frac{-0.284 \times 10^{-3} - (-0.798 \times 10^{-3})}{3(13)} = 1.32 \times 10^{-5}$$

$$d_2 = \frac{-0.301 \times 10^{-3} - (-0.284 \times 10^{-3})}{3(15)} = -3.787 \times 10^{-7}$$

$$d_3 = \frac{0 - (-0.301 \times 10^{-3})}{3(17)} = 5.90 \times 10^{-6}$$

Similarly, substituting into (4.30), we get

$$b_0 = \frac{1}{11} - \frac{11}{3}\left(-0.798 \times 10^{-3}\right) = 0.0938$$

$$b_1 = \frac{1}{13} - \frac{13}{3}\left(-0.284 \times 10^{-3} + 2\left(-0.798 \times 10^{-3}\right)\right) = 0.0851$$

$$b_2 = \frac{1}{15} - \frac{15}{3}\left(-0.301 \times 10^{-3} + 2\left(-0.284 \times 10^{-3}\right)\right) = 0.0710$$

$$b_3 = \frac{1}{17} - \frac{17}{3}\left(0 + 2\left(-0.301 \times 10^{-3}\right)\right) = 0.0622$$

The components of the natural cubic interpolation spline for this data are then

$$s_k(x) = a_k + b_k(x - x_k) + c_k(x - x_k)^2 + d_k(x - x_k)^3$$

So, for example, with $x = 30 \in [25, 36]$, we use $k = 0$ and $x - x_0 = 5$ so that

$$s(30) = 5 + (0.0938)(5) + 0(5)^2 + \left(-2.22 \times 10^{-5}\right)(5)^3$$
$$= 5.4662$$

Similarly, for $x = 40 \in [36, 49]$, we use $k = 1$, $x - x_1 = 4$ and obtain

$$s(40) = 6 + (0.0851)(4) + \left(-0.798 \times 10^{-3}\right)(4)^2 + \left(1.32 \times 10^{-5}\right)(4)^3$$
$$= 6.3285$$

The errors in the final values in Example 17 are approximately 0.011 and 0.004, which are of a similar order of magnitude to those that would be expected from using local cubic divided difference interpolation. (The data in Example 17 are, of course, taken from the square-root function.) This is to be expected since the error in natural cubic spline interpolation is again of order h^4 as it is for the cubic polynomial interpolation. The proof of this result for splines is much more difficult, however. It is left to more advanced texts.

Typically, the natural cubic spline will have better smoothness properties because, although the approximating polynomial pieces are local, they are affected by the distant data. In order to get a good match to a particular curve, we must be careful about the placement of the knots. We shall consider this further, though briefly, after discussing the implementation of natural cubic spline interpolation.

The m-file cspline.m first computes the coefficients using the equations derived earlier. The tridiagonal system is solved here using MATLAB's built-in linear equation solving operator \. We shall discuss techniques for solving such systems in detail in Chapter 7. For now we shall content ourselves with the 'black box' provided with MATLAB.

Program	MATLAB m-file for cubic spline interpolation

```
function s=cspline(knots,data,x);
% Computes coefficients for
% Natural Cubic Spline Interpolation
% at 'knots' with values 'data'
% and evaluates the spline at all points of the array x
N=length(knots)-1;
P=length(x);
h=diff(knots); % computes 'steplength' vector
D=diff(data)./h; % computes first divided differences of data
dD3=3*diff(D); % differences of D for rhs of linear system
a=data(1:N);
% Next generate tridiagonal system
H=diag(2*(h(2:N)+h(1:N-1)));
for k=1:N-2
  H(k,k+1)=h(k+1);
  H(k+1,k)=h(k+1);
end
c=zeros(1,N+1); % initializes and fixes 'normal' end conditions
c(2:N)=H \ dD3';
b=D-h.*(c(2:N+1)+2*c(1:N))/3;
d=(c(2:N+1)-c(1:N))./(3*h);   % completes computation of coefficients
% Begin evaluation of spline at x-values
for i=1:P
  k=1;
  while (x(i)>knots(k+1)) & (k<N)
    k=k+1;
  end; % x(i) lies in knot interval [knots(k), knots(k+1)]
  z=x(i)-knots(k);
  s(i)=a(k)+z*(b(k)+z*(c(k)+z*d(k)));
end;
```

Note the use of array arithmetic operations in generating the remaining coefficients once the cs have been computed. Note, too, the use of diag to initialize the matrix elements.

This m-file assumes that the arguments x are all in the interval spanned by the knots. It would evaluate the spline outside this interval by using the first component for any x-value to the left, and the last component for values to the right of the span of the knots. It is not difficult to modify the code to extend the spline with straight line components beyond this interval but we do not trouble with this here.

Example 15   **Plot the error function $s(x) - \sqrt{x}$ for the data of Example 14**

We can use the m-file cspline.m with the following commands to generate the graph below.

```
» X=[25 36 49 64 81]; Y=[5 6 7 8 9];
» x=[25:0.4:81];
» y=cspline(X,Y,x);
» plot(x,y-sqrt(x))
```

Here the vectors X,Y are the knots and the corresponding data values, x is a vector of points at which to evaluate the natural cubic spline interpolating this data and y is the corresponding vector of spline values (Figure 4.7).

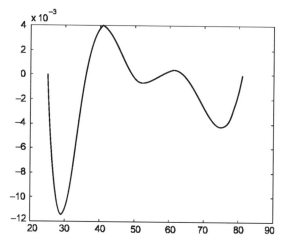

Figure 4.7   Error in cubic spline interpolation

We see that this error function varies quite smoothly, as we hoped. Also we note that the most severe error occurs very near $x = 30$, which was the first point we used in the earlier example. The evidence presented there was representative of the overall performance.

In motivating the study of spline interpolation, we used an example of local cubic divided difference interpolation where the resulting piecewise polynomial had a discontinuity (see Figure 4.5). The corresponding cubic spline interpolant is plotted in Figure 4.8 below.

This was generated using the MATLAB commands

```
» X=[0,1:0.25:2,3]; Y=[1,log(X(2:7))];
» y=cspline(X,Y,x);
» plot(x,y,X,Y,'k*')
```

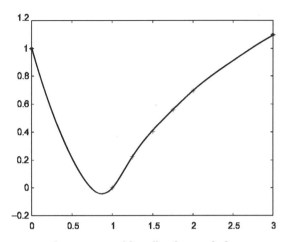

Figure 4.8   Cubic spline interpolation

It is immediately apparent that the cubic spline has handled the changes in this curve much more readily. The main cause of difficulty for the local polynomial interpolation used earlier is the fact that the function has a minimum and an inflection point very close together – and the inflection point appears to be very close to the discontinuity of the local divided difference interpolation function.

In general, we want to place more nodes in regions where the function is changing rapidly (especially in its first derivative). We shall consider briefly the question of knot placement for cubic spline interpolation in the next example.

Example 16   Use cubic spline interpolation for data from the function $1/(1 + x^2)$ with eleven knots in $[-5, 5]$

First, for equally spaced knots, we can use the MATLAB commands

```
» X=-5:5;
» Y=runge(X);
» x=-5:0.05:5;
» plot(X,Y,'*',x,runge(x),x,cspline(X,Y,x))
» plot(x,runge(x)-cspline(X,Y,x))
```

to obtain the plots in Figures 4.9 and 4.10.

In Figure 4.9, we see that the graphs of Runge and its cubic spline interpolant using integer knots in $[-5, 5]$ are essentially indistinguishable. The error function shows that the fit is especially good near the ends of the range (where polynomial interpolation failed). The errors are largest either side of ± 1 which is where the inflection points are, and where the gradient is changing rapidly. To get a more uniform fit we would probably want to place more of the knots in this region.

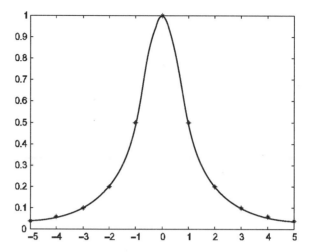

Figure 4.9   Cubic spline interpolation of Runge

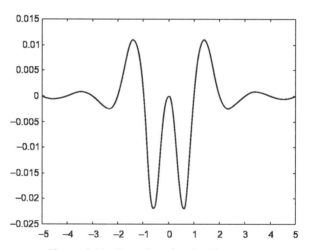

Figure 4.10   Error function for Figure 4.9

In Figure 4.11, we see the effect on the error function of using knots at

$$-5, \ -3.5, \ -2.5, \ -1.5, \ -0.75, \ 0, \ 0.75, \ 1.5, \ 2.5, \ 3.5, \ 5$$

There are still just 11 knots but the magnitude of the maximum error has been reduced by an approximate factor of 10.

This set of knots is fairly easily generated in MATLAB as follows:

```
» X=[0.75,1.5,2.5,3.5,5];
» X=[fliplr(-X),0,X];
```

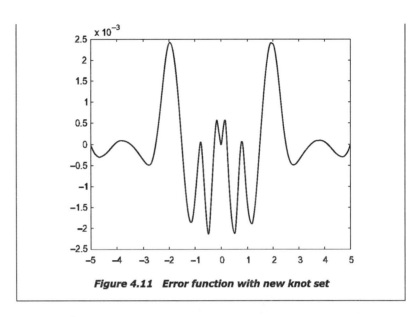

Figure 4.11   Error function with new knot set

We have seen that natural cubic splines can be used to obtain very smooth interpolating functions even for data where polynomial interpolation cannot reproduce the correct behaviour.

Splines are a very powerful tool. They are used as the basis of much computer aided design (CAD). In particular, Bezier splines were developed originally for the French automobile industry in the early 1960s for this specific purpose. Bezier splines are a generalization of the cubic splines we have seen to general planar curves. They are not necessarily interpolation splines but make us of what are called 'control points' which are used to 'pull' a curve in a particular direction. Many computer graphics programs use similar techniques for generating smooth curves through particular points. Splines were, for example, the primary mathematical tool used by Pixar in producing their computer animation masterpieces *Toy Story* and *A Bug's Life*.

Exercises: Section 4.4

1   Which of the following functions are splines of degree 1, 2, or 3?

(a) $s(x) = \begin{cases} x & 0 \leq x \leq 1 \\ 2x - 1 & 1 \leq x \leq 2 \\ x + 2 & 2 \leq x \leq 4 \end{cases}$

(b) $s(x) = \begin{cases} 2 - x & 0 \leq x \leq 1 \\ 2x - 1 & 1 \leq x \leq 2 \\ x + 1 & 2 \leq x \leq 4 \end{cases}$

(c) $s(x) = \begin{cases} x^2 & -1 \leq x \leq 1 \\ 2x^2 - 2x + 1 & 1 \leq x \leq 2 \\ 3x^2 - 6x + 5 & 2 \leq x \leq 3 \end{cases}$

(d) $s(x) = \begin{cases} x^3 & 0 \leq x \leq 1 \\ 2x^2 - 2x + 1 & 1 \leq x \leq 2 \\ 3x^2 - 6x + 5 & 2 \leq x \leq 4 \end{cases}$

(e) $s(x) = \begin{cases} x & -1 \leq x \leq 0 \\ x + x^3 & 0 \leq x \leq 1 \\ 1 - 2x + 3x^2 & 1 \leq x \leq 4 \end{cases}$

2   Find the linear spline which interpolates the data

x	0	1	3	4	6
$f(x)$	5	4	3	2	1

What are its values at 2, 3.5 and 4.5?

3   Write a MATLAB m-file to compute the linear spline interpolating a given set of data. Use it to plot the linear spline interpolant for the data in Exercise 2. Also plot the linear interpolating spline for the function $1/(1 + x^2)$ using knots $-5, -4, \ldots, 5$. (If *all* we want is the plot, there is a simpler way, of course. Just plot(X,Y) where X and Y are the vectors of knots and their corresponding data values.)

4   Find the natural cubic interpolation spline for the data

x	1	2	3	4	5
$\ln x$	0.0000	0.6931	1.0986	1.3863	1.6094

(**Hints:** As usual, the a-coefficients are just the data values:

$$a_0 = 0.0000 \quad a_1 = 0.6931 \quad a_2 = 1.0986 \quad a_3 = 1.3863$$

You should obtain the tridiagonal system for c_1, c_2, c_3:

$$\begin{bmatrix} 4 & 1 & 0 \\ 1 & 4 & 1 \\ 0 & 1 & 4 \end{bmatrix} \begin{bmatrix} c_1 \\ c_2 \\ c_3 \end{bmatrix} = \begin{bmatrix} -0.8628 \\ -0.3534 \\ -0.1938 \end{bmatrix})$$

5   Compute the natural cubic spline interpolant for data from the function $\sqrt{x}\exp(-x)$ using knots at the integers in $[0, 8]$. Compare the graph of this spline with that of $\sqrt{x}\exp(-x)$.

6   We can continue a natural cubic spline outside the span of the knots with straight lines while preserving the continuity of the first 2 derivatives. Show that to do this we can use $s_{-1}(x) = a_0 + b_0(x - x_0)$ and $s_n(x) = a_n + b_n(x - x_n)$ where

$$a_n = f(x_n), \quad \text{and} \quad b_n = \frac{c_{n-1}h_{n-1}}{3} + \delta_{n-1}$$

7   Modify the m-file for natural cubic spline interpolation to include straight line extensions outside the span of the knots. Use this modified code to plot the natural cubic spline interpolant to $1/(1 + x^2)$ over the interval $[-6, 6]$ using knots $-5, -4, \ldots, 5$.

8   Repeat Exercise 5, using 17 equally spaced knots, 0:0.5:8.

9   Try to find the best set of 11 knots for natural cubic spline interpolation to $1/(1 + x^2)$ over $[-5, 5]$

4.5 MATLAB interpolation functions

MATLAB has several built-in functions related to interpolation. These cover both polynomial and spline interpolation.

interp1   is MATLAB's one-dimensional interpolation function. It takes arguments consisting of a vector of nodes, a vector of corresponding data values and the vector of points at which the interpolating function is to be evaluated. There is a further optional argument which is the type of interpolation to be used.

The options for the fourth argument are:

linear, which is also the default, which performs linear interpolation between the data values. This is equivalent to linear spline interpolation.

cubic, which uses local cubic polynomial interpolation similar to the local divided difference scheme discussed in Section 4.3.

spline, which performs cubic spline interpolation by an algorithm similar to that of Section 4.4.

nearest, which is a piecewise *constant* interpolation where the values returned are simply the data values at the nearest node. This is equivalent to a spline of degree 0.

Example 17   Using the data

  » X=[0,1:0.25:2,3]; Y=[1,log(X(2:7))];

The various options in MATLAB's interp1 m-file produce the following graphs.

  » x=[0:0.02:3];
  » y1=interp1(X,Y,x);
  » y2=interp1(X,Y,x,'cubic');
  » y3=interp1(X,Y,x,'spline');
  » y4=interp1(X,Y,x,'nearest');
  » plot(x,y1,x,y2,x,y3,x,y4)

(See Figure 4.12.)

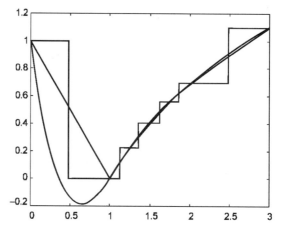

Figure 4.12   Using MATLAB's interp1 *function*

In this case we see there is no discernible difference between the local cubic interpolation and the cubic spline implementation. The details of the spline implementation are different from those of Section 4.4 because MATLAB's spline is not necessarily the natural cubic spline. In Figure 4.13, the result of plotting our cspline function is included. We see that the natural cubic spline does not dip as low as the MATLAB spline in making the turn but then produces essentially the same curve beyond $x = 1$ (Figure 4.13).

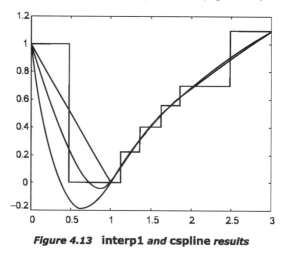

Figure 4.13   interp1 *and* cspline *results*

MATLAB also has functions interp2 and interpn for performing two-dimensional, and n-dimensional, interpolation. These are beyond our present scope and the reader is referred to the MATLAB manuals or help files for further information on these.

spline is another MATLAB function. Its effect is precisely the same as the 'spline' option in interp1. One difference is that this function can be used with just the knots and data arguments to obtain the piecewise polynomial itself which can be used later with ppval for evaluation or plotting. For details of this see the MATLAB manual.

For more advanced applications of spline interpolation and approximation, MATLAB has the specialist Spline Toolbox which is a separately purchased add-on to the basic MATLAB package.

polyfit is really designed for *least squares fitting* of polynomials to data. It generates the polynomial (given by its coefficients) which best fits a given set of data points. One of the arguments is the degree of the polynomial to be used.

For full polynomial interpolation in MATLAB we can use polyfit with the degree set to one less than the number of nodes. In this case it will generate the Lagrange interpolation polynomial by solving the Vandermonde system. To evaluate the resulting polynomial we then need to use the function polyval.

Example 18 Use the same data as in Example 17 with polyfit to obtain the Lagrange interpolation polynomial

There are 7 nodes and so we seek a polynomial of degree 6. The commands

```
» p=polyfit(X,Y,6);
» y=polyval(p,x);
» plot(x,y)
```

compute this interpolation polynomial and the plot in Figure 4.14.

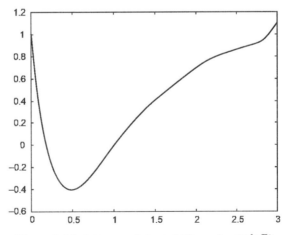

Figure 4.14 Lagrange interpolation using polyfit

It is immediately apparent that this polynomial is behaving significantly differently at the right-hand end of the range. Try extending the range of the plot to see just how different it becomes.

We shall return to the use of polyfit for least squares polynomial fitting in Chapter 7. To give a foretaste of just how different such an approximation may be, in Figure 4.15, we have plotted the Lagrange interpolation polynomial along with the least squares cubic fit to the same data.

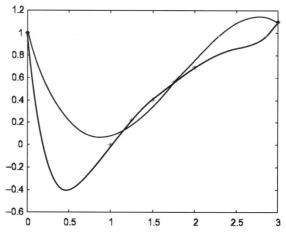

Figure 4.15   Least squares cubic approximation

This is computed by

```
» p3=polyfit(X,Y,3);
» y3=polyval(p3,x);
```

5 Numerical Calculus

Aims and objectives

In this chapter, we move beyond the evaluation of a function to approximations for the fundamental operations of the calculus: *numerical differentiation, integration* and *optimization*. We see that *numerical integration* is much more reliable than is numerical differentiation. There are many efficient techniques for the former, and we introduce the basic ideas of these, again blending the theoretical and the practical. All of the tasks considered have their foundations in interpolation, providing further evidence for the value of that theory. As usual, the primary objective is to give the reader a good feel for the basic ideas.

5.1 Introduction

Chapters 3 and 4 were concerned with numerical techniques for evaluation or approximation of functions. We have discussed a number of different approaches to this problem. However, the evaluation of a function is but the beginning; many other questions arise in order that we might understand the nature of any particular function. Familiar questions of the calculus are those of differentiation, integration and optimization of functions.

In this chapter, we consider some of the numerical methods of answering these questions. The methods we discuss here are, for the most part, derived from the ideas of interpolation. Thus, for example, the basic numerical integration formulas which we consider can be obtained by integrating interpolation polynomials. Similarly, the numerical differentiation formulas result from differentiation of interpolation polynomials, although they are easier to obtain by considering the relation between derivatives and divided differences. The quadratic and cubic search algorithms for minimization (or maximization) of a function are also based on applying simple calculus techniques to (sequences of) interpolating quadratic or cubic polynomials.

Before discussing any of these topics in detail, we consider the question of why we need approximate methods for tasks for which there are perfectly well understood and elementary calculus techniques already available.

Example 1 The need for numerical integration

Consider the evaluation of the standard normal distribution function of statistics

$$N_{0,1}(x) = \frac{1}{\sqrt{2\pi}} \int_{-\infty}^{x} \exp\left(-t^2/2\right) dt = \frac{1}{2} + \frac{1}{\sqrt{2\pi}} \int_{0}^{x} \exp\left(-t^2/2\right) dt \quad (5.1)$$

This integral is of vital importance for statistical analysis, but it *cannot* be evaluated by conventional calculus techniques – because there is no antiderivative for $e^{-x^2/2}$ expressible in terms of elementary functions. Of course, we can (and do) *define* a function *erf* similar to this antiderivative but that just gives it a name; it does not help with its evaluation. To evaluate this function, we need approximation methods. One possibility is to use numerical integration methods similar to those discussed in this chapter. (There are other possibilities based on series expansions which can be used for this particular integral.) We shall return to this example later.

Most arc length problems result in integrals which cannot be evaluated exactly: integrals of the form

$$L = \int_{a}^{b} \sqrt{1 + (f'(x))^2} \, dx$$

for the arc length of the curve $y = f(x)$ between $x = a$ and $x = b$ are rarely doable by calculus methods. The need to be able to evaluate arc lengths, and similar, though more complicated, integrals for surface areas is apparent.

The Riemann integral is defined as a limit of Riemann sums for arbitrary partitions of the range of integration. This is usually approximated, at an elementary level, by taking *uniform* partitions with some number N of subdivisions. We can then denote the *steplength* of these subdivisions by

$$h = \frac{b-a}{N}$$

and write

$$x_k = a + kh \qquad (k = 0, 1, \ldots, N)$$

Then the Riemann left and right sums (Figure 5.1) are given by

$$L = h \sum_{k=0}^{N-1} f(x_k) = h \sum_{k=0}^{N-1} f(a + kh) \tag{5.2}$$

$$R = h \sum_{k=1}^{N} f(x_k) = h \sum_{k=1}^{N} f(a + kh) \tag{5.3}$$

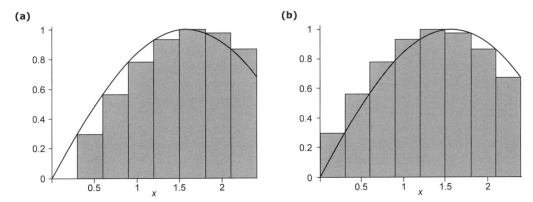

Figure 5.1   Left and right sum approximations to an integral

These Riemann sums can be viewed as very crude numerical integration techniques. They are equivalent to integrating step-functions which agree with f at the appropriate set of nodes. These step-functions are piecewise degree 0 interpolating polynomials. The efficient methods we shall investigate result from increasing this degree.

Step function approximations to an integral are illustrated in Figure 5.1. It is easy to see that the shaded areas will differ substantially from the true value of the integral. Clearly there is a need for better numerical integration methods than these.

Numerical differentiation becomes necessary in circumstances where the actual derivative is not available. Such situations arise in, for example, shooting methods for solving boundary value problems. We shall discuss these in some detail in Chapter 6. If a function is available only from a table of values, perhaps from some experimental data, then we have no access to its exact derivative. However, it may be necessary to estimate this derivative. Another situation in which numerical estimates of derivatives are widely used is in finite difference solutions of differential equations.

We do not typically confront the situation where a function is known by a formula and yet we cannot differentiate it. The rules of differentiation allow us to differentiate a function which is composed of more elementary functions which we can differentiate. This is the basic reason why symbolic differentiation is relatively easy, while symbolic integration is not. The situation is precisely reversed for numerical methods. Efficient and accurate numerical integration is much easier than differentiation.

Even for functions of a single variable, the standard calculus technique for optimization of testing for critical points is not always sufficient. It is easy to find examples where setting the derivative to zero results in an equation which we cannot solve algebraically. Applying our equation solving techniques may help – but will also ignore much of the information that can be obtained from the function itself. This function information can be used to avoid locating local maxima while we seek the minimum of the objective function, for example. We consider the question of locating local maxima and minima later in this chapter.

5.2 Numerical integration: interpolatory quadrature rules

In this section, we introduce the ideas of numerical integration, in which we approximate a definite integral

$$I = \int_a^b f(x)dx$$

by a weighted sum of values of the function f at some finite set of points. That is, we use approximations of the form

$$\int_a^b f(x)dx \approx \sum_{k=0}^{N} c_k f(x_k) \tag{5.4}$$

where the *nodes* $x_0, x_1, \ldots, x_N \in [a, b]$ and the coefficients, or *weights*, c_k are chosen appropriately.

A formula such as (5.4) is a special form of a Riemann sum where the nodes are the representative points from the various intervals in the partition. The weights represent the widths of the subdivisions of the interval. Although such formulas *are* Riemann sums, this does not suggest a useful way of obtaining numerical integration, or *quadrature*, rules.

The aim of a good numerical integration algorithm is that its formula should be both simple and efficient – and, of course, it should provide good approximations of the true integrals. All the methods we shall consider are based on integrating a polynomial which has the same values as f at the nodes – that is, an interpolation polynomial.

The fact that such integration formulas are derived from integrating interpolation polynomials also provides us with a method of finding the weights in (5.4). Suppose p is the interpolation polynomial agreeing with f at the nodes $x_0, x_1, \ldots, x_N$. Then the corresponding *interpolatory quadrature*, or *numerical integration rule*, is given by

$$\int_a^b f(x)dx \approx \int_a^b p(x)dx$$

$$= \int_a^b \sum_{k=0}^{N} f(x_k)l_k(x)dx$$

$$= \sum_{k=0}^{N} c_k f(x_k) \tag{5.5}$$

where the $l_k(x)$ are the Lagrange basis polynomials which we met in Chapter 4. It follows, therefore, that

$$c_k = \int_a^b l_k(x)dx \qquad (k = 0, 1, \ldots, N) \tag{5.6}$$

Again, however, this does *not* provide a good general technique for obtaining these weights.

Now, if the integrand f were itself a polynomial of degree not more than N, then the interpolation polynomial p would be f itself. Then integrating the interpolation polynomial would yield the exact integral. This consideration leads us to the following definition.

Definition 1 The numerical integration formula (5.4) has *degree of precision m* if it is exact for all polynomials of degree at most m, but is not exact for x^{m+1}.

Any interpolatory quadrature rule using the nodes x_0, x_1, ..., x_N must have degree of precision *at least* N since it necessarily integrates polynomials of degree up to N exactly. (We shall see later that the degree of precision can be greater than N.) This observation provides us with a more amenable technique for obtaining the weights than finding and integrating all the Lagrange basis polynomials. We illustrate the process with an example.

Example 2 Find the interpolatory quadrature rule using the nodes $x_0 = 0, x_1 = 1$, and $x_2 = 3$ for approximating $\int_0^4 f(x)dx$. Use the resulting rule to estimate $\int_0^4 x^3 dx$

Using three nodes, we expect a degree of precision of at least 2. Thus, we seek weights c_0, c_1, c_2 such that

$$I = \int_0^4 f(x)dx = c_0 f(0) + c_1 f(1) + c_2 f(3)$$

for $f(x) = 1, x, x^2$. Now, with

$$f(x) \equiv 1 \quad I = 4 \qquad c_0 f(0) + c_1 f(1) + c_2 f(3) = c_0 + c_1 + c_2$$
$$f(x) = x \quad I = 8 \qquad c_0 f(0) + c_1 f(1) + c_2 f(3) = c_1 + 3c_2$$
$$f(x) = x^2 \quad I = 64/3 \quad c_0 f(0) + c_1 f(1) + c_2 f(3) = c_1 + 9c_2$$

Solving the equations

$$c_0 + c_1 + c_2 = 4, \quad c_1 + 3c_2 = 8, \quad c_1 + 9c_2 = 64/3$$

we obtain

$$c_0 = \frac{4}{9}, \quad c_1 = \frac{4}{3}, \quad c_2 = \frac{20}{9}$$

The required quadrature rule is therefore

$$\int_0^4 f(x)dx \approx \frac{4f(0) + 12f(1) + 20f(3)}{9}$$

Now, with $f(x) = x^3$, we get

$$\int_0^4 x^3 dx \approx \frac{4(0)^3 + 12(1)^3 + 20(3)^3}{9} = \frac{184}{3}$$

whereas the true integral is $4^4/4 = 64$. This shows that the degree of precision of this formula is 2 since it fails to be exact for x^3. Also we see that the error for $f(x) = x^3$ is only 8/3, or approximately 4%.

Fortunately, it is not untypical of numerical integration that small relative errors can be obtained from simple approaches. Clearly, though, the distribution of nodes will affect the accuracy that can be obtained. In Example 2 there is only 1 node in the right-hand three-quarters of the range of integration. In order to obtain greater accuracy, we will typically require more points, and they will need to be more evenly spread across the range of integration.

In Example 2, we set out to find a quadrature rule with degree of precision at least 2, and the procedure was to find weights so that the rule is exact for the functions 1, x, x^2. The fact that this is sufficient to establish exactness for all quadratics follows readily from the linearity of both the integral and the numerical integration formula. Thus, for any quadratic polynomial $p(x) = \alpha x^2 + \beta x + \gamma$, we have

$$\int_a^b p(x)dx = \alpha \int_a^b x^2 dx + \beta \int_a^b x dx + \gamma \int_a^b 1 dx$$

and a similar result holds for the quadrature rule.

It is clear that, following the same procedure used in Example 2, we can find the weights for the interpolatory quadrature rule using any set of nodes by solving a system of linear equations. We shall concentrate initially on just three elementary integration formulas which form the basis of some powerful and practical techniques for numerical integration.

The midpoint rule (Figure 5.2)

$$\int_a^b f(x)dx \approx (b-a)f((a+b)/2) \tag{5.7}$$

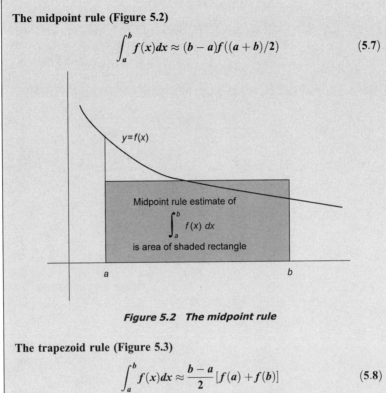

Figure 5.2   *The midpoint rule*

The trapezoid rule (Figure 5.3)

$$\int_a^b f(x)dx \approx \frac{b-a}{2}\left[f(a)+f(b)\right] \tag{5.8}$$

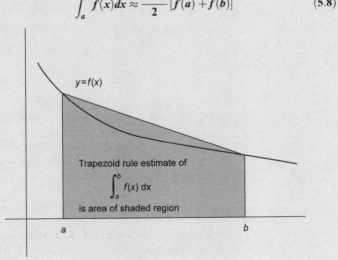

Figure 5.3   *The trapezoid rule*

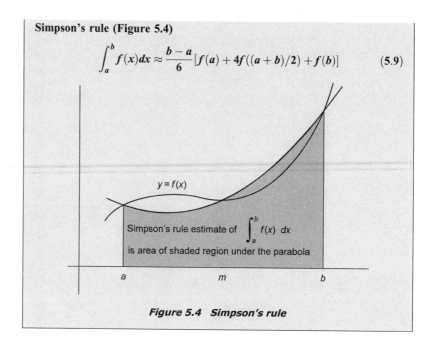

Simpson's rule (Figure 5.4)

$$\int_a^b f(x)dx \approx \frac{b-a}{6}[f(a) + 4f((a+b)/2) + f(b)] \qquad (5.9)$$

$y = f(x)$

Simpson's rule estimate of $\int_a^b f(x)\ dx$

is area of shaded region under the parabola

$a \qquad\qquad m \qquad\qquad b$

Figure 5.4 Simpson's rule

These can be obtained by integrating the interpolation polynomials which agree with f at the given nodes, or by solving the appropriate linear systems. We shall illustrate the derivation with the special case of Simpson's rule for the interval $[-h, h]$.

We require an integration rule using the nodes $-h$, 0, h which is exact for all quadratics. Therefore, we must find coefficients α, β, γ such that

$$I = \int_{-h}^h f(x)dx = \alpha f(-h) + \beta f(0) + \gamma f(h)$$

for $f(x) = 1, x, x^2$. These requirements yield the equations

$$\alpha + \beta + \gamma = 2h$$
$$-\alpha h + \gamma h = 0$$
$$\alpha(-h)^2 + \gamma h^2 = \frac{2h^3}{3}$$

The second of these immediately gives us $\alpha = \gamma$, and then the third yields $\alpha = h/3$. Substituting these into the first equation, we get $\beta = 4h/3$ so that Simpson's rule for this interval is

$$\int_{-h}^h f(x)dx \approx \frac{h}{3}[f(-h) + 4f(0) + f(h)]$$

For a general interval $[a, b]$, we can use $h = (b - a)/2$ and $m = (a + b)/2$ with the change of variables

$$x = \frac{a + b}{2} + s$$

to get

$$\int_a^b f(x)dx = \int_{-h}^{h} f(m + s)ds \approx \frac{h}{3}[f(a) + 4f(m) + f(b)]$$

which corresponds to (5.9).

The derivations of the midpoint and trapezoid rules are similar – but simpler. They are left as exercises.

We remarked earlier that it is possible for a quadrature rule to have a higher degree of precision than that of the interpolation polynomial. Simpson's rule is a case in point; it is exact not just for quadratics but for cubics, too. To see this consider applying Simpson's rule to $\int_{-h}^{h} x^3 dx$. The integral evaluates to 0, since it is an integral of an odd function over a symmetric interval. Simpson's rule gives the approximation

$$\frac{h}{3}\left[(-h)^3 + 4(0)^3 + h^3\right] = -\frac{h^4}{3} + \frac{h^4}{3} = 0$$

which is exact, as claimed. That its degree of precision is 3 and not higher follows from the fact that Simpson's rule is not exact for $f(x) = x^4$.

The midpoint rule gives a similar bonus: it has degree of precision 1. Some numerical integration rules therefore provide proof that there really is such a thing as a (mathematical) free lunch!

The performance of these rules for low-degree polynomials is not our primary concern. We need to get some idea of how well, or otherwise, they will approximate more general integrals. Error formulas for interpolatory quadrature rules can be obtained by integrating the Lagrange interpolation remainder term. Only very rarely is this process other than *extremely* messy, and we shall omit the details entirely. Later, we shall, however, give some reasons to believe the results quoted below for the errors in the midpoint, trapezoid and Simpson's rules:

$$\int_a^b f(x)dx - (b - a)f((a + b)/2) = \frac{(b - a)h^2}{24}f''(\xi_M) \qquad (5.10)$$

where $h = b - a$,

$$\int_a^b f(x)dx - \frac{b - a}{2}[f(a) + f(b)] = -\frac{(b - a)h^2}{12}f''(\xi_T) \qquad (5.11)$$

where again $h = b - a$, and

$$\int_a^b f(x)dx - \frac{h}{3}[f(a) + 4f(a + h) + f(b)] = -\frac{(b - a)h^4}{180}f^{(4)}(\xi_S) \qquad (5.12)$$

where $h = (b - a)/2$. Here the points ξ_M, ξ_T, ξ_S are 'mean value points' in the interval (a, b).

The reason for the, apparently unnecessary, use of $h = b - a$ in (5.10) and (5.11) will become apparent in the next section when we look at the 'composite' forms of these integration rules.

Example 3 **Estimate**

$$N_{0,1}(1/2) = \frac{1}{2} + \frac{1}{\sqrt{2\pi}} \int_0^{1/2} \exp\left(-x^2/2\right) dx$$

(a) Using the midpoint rule:

$$N_{0,1}(1/2) \approx \frac{1}{2} + \frac{1}{\sqrt{2\pi}} \frac{1}{2} \exp\left(-(1/4)^2/2\right) \approx 0.693334$$

(b) Using the trapezoid rule

$$N_{0,1}(1/2) \approx \frac{1}{2} + \frac{1}{\sqrt{2\pi}} \frac{1/2}{2} \left[\exp(0) + \exp\left(-(1/2)^2/2\right)\right] \approx 0.687752$$

(c) Using Simpson's rule

$$N_{0,1}(1/2) \approx \frac{1}{2} + \frac{1}{\sqrt{2\pi}} \frac{1/4}{3} \left[\exp(0) + 4\exp\left(-(1/4)^2/2\right) + \exp\left(-(1/2)^2/2\right)\right]$$

$$\approx 0.691473$$

The true value is 0.6914625 to 7 decimals. The errors in these approximations are therefore approximately $0.0019, 0.0037$ and 0.00001, respectively. It is obvious that Simpson's rule has given a significantly better result. We also see that the error in the midpoint rule is close to one-half that of the trapezoid rule as suggested by the error formulas (5.10) and (5.11).

We have seen that interpolatory quadrature rules can have degree of precision greater than we should expect from the number of nodes. If we are free to choose the nodes themselves, as well as the appropriate weights attached to them, then it is worthwhile to seek the quadrature rule with maximum degree of precision for a given number of nodes. Such an approach to numerical integration is known as *Gaussian quadrature*. The general theory of Gaussian integration is beyond the scope of the present text. However, we can get a taste of its rich theory through some examples.

Example 4 **Find the interpolatory quadrature formulas with maximum degree of precision, the Gaussian rules, on $[-1, 1]$ using**

(a) **two points, and**

(b) **three points**

For simplicity, we shall denote $\int_{-1}^{1} f(x)dx$ by $I(f)$.

(a) We must find the nodes x_0, x_1 and the associated weights c_0, c_1 to make the integration rule exact for $f(x) = 1, x, x^2, \ldots$ as far as we can go. Since there are 4 unknowns, it is reasonable to try to satisfy 4 such equations. We recall that

$$I(x^k) = \int_{-1}^{1} x^k dx = \begin{cases} \dfrac{2}{k+1} & \text{if } k \text{ is even} \\ 0 & \text{if } k \text{ is odd} \end{cases}$$

Thus we require:

(i)   For $f(x) = 1$, $I(f) = 2 = c_0 + c_1$
(ii) For $f(x) = x$, $I(f) = 0 = c_0 x_0 + c_1 x_1$
(iii) For $f(x) = x^2$, $I(f) = \frac{2}{3} = c_0 x_0^2 + c_1 x_1^2$
(iv) For $f(x) = x^3$, $I(f) = 0 = c_0 x_0^3 + c_1 x_1^3$

We can assume that neither c_0 nor c_1 is zero since we would then have just a one-point formula. From (ii), we see that $c_0 x_0 = -c_1 x_1$, and substituting this in (iv), we get

$$0 = c_0 x_0^3 + c_1 x_1^3 = c_0 x_0 \left(x_0^2 - x_1^2 \right)$$

We have already concluded that $c_0 \neq 0$. If $x_0 = 0$, then (ii) would also give us $x_1 = 0$ and again the formula would reduce to the midpoint rule. It follows that $x_0^2 = x_1^2$, and therefore that $x_0 = -x_1$. Now (ii) implies that $c_0 = c_1$, and from (i), we deduce that $c_0 = c_1 = 1$. Substituting all of this into (iii), we get $2x_0^2 = 2/3$ so that we can take

$$x_0 = -\frac{1}{\sqrt{3}}, \quad x_1 = \frac{1}{\sqrt{3}}$$

The resulting integration rule is

$$\int_{-1}^{1} f(x)dx \approx f\left(-1/\sqrt{3}\right) + f\left(1/\sqrt{3}\right) \tag{5.20}$$

Note that, by appealing to the symmetry of the situation, we could immediately deduce that $x_0 = -x_1$ and $c_0 = c_1$, which would shorten this derivation considerably.

(b) This time, we seek three nodes x_0, x_1, x_2 and their associated weights c_0, c_1, c_2. This time, we shall simplify the argument by using symmetry. If the nodes are ordered in their natural order, then we may deduce from symmetry that

$$x_0 = -x_2, \quad x_1 = 0$$
$$c_0 = c_2$$

Using these facts in the equations for $f(x) = 1, x^2, x^4$, we get

$$2c_0 + c_1 = 2$$

$$2c_0 x_0^2 = \frac{2}{3}$$

$$2c_0 x_0^4 = \frac{2}{5}$$

(The equations for $f(x) = x, x^3, x^5$ are automatically satisfied with $x_0 = -x_2, x_1 = 0$ and $c_0 = c_2$.) The last pair of these yield $x_0^2 = 3/5$ and then $c_0 = 5/9$, from which we get $c_1 = 8/9$. The Gaussian quadrature using three nodes in $[-1, 1]$ is therefore

$$\int_{-1}^{1} f(x)dx \approx \frac{1}{9}\left[5f\left(-\sqrt{3/5}\right) + 8f(0) + 5f\left(\sqrt{3/5}\right)\right] \qquad (5.21)$$

and this has degree of precision 5.

Example 5   Use the Gaussian rules with 2 and 3 nodes to estimate $\int_{-1}^{1} 1/(1+x^2)dx$; compare the results with those obtained using the midpoint, trapezoid and Simpson's rules

For comparison, we first obtain the midpoint, trapezoid and Simpson's rule estimates:

$$M = 2\left(\frac{1}{1+0^2}\right) = 2$$

$$T = 1\left[\frac{1}{1+(-1)^2} + \frac{1}{1+1^2}\right] = 1$$

$$S = \frac{1}{3}\left[\frac{1}{2} + 4(1) + \frac{1}{2}\right] = \frac{5}{3} \approx 1.66667$$

The true result is $\pi/2 \approx 1.57080$.

The 2 Gaussian rules obtained in Example 4 yield

$$(5.13) \quad G_2 = \frac{1}{1+\left(-1/\sqrt{3}\right)^2} + \frac{1}{1+\left(1/\sqrt{3}\right)^2} = 1.5$$

$$(5.14) \quad G_3 = \frac{1}{9}\left[\frac{5}{1+3/5} + 8 + \frac{5}{1+3/5}\right] = \frac{114}{72} \approx 1.58333$$

It is apparent that the Gaussian rules have significantly greater accuracy than the simpler rules using similar numbers of nodes.

Unfortunately, this Gaussian quadrature approach does not lend itself readily to use within an automatic numerical integration scheme, since the nodes and weights for one rule are not related in a simple way to those of other Gaussian rules. There has been considerable effort devoted to the development of efficient quadrature routines based on Gaussian quadrature but we do not pursue these here. Some of the most important properties of Gaussian quadrature on the interval $[-1, 1]$ are summarized in the following theorem whose proof is omitted.

Theorem 1 *(Gaussian quadrature rules on $[-1, 1]$)*

The Gaussian quadrature rule using N nodes in $[-1, 1]$ has degree of precision $2N - 1$. Its nodes are the roots of the Legendre polynomial of degree N, and the weights are all positive

---

The Legendre polynomials are a special system of polynomials, called *orthogonal polynomials*, which are given by the recurrence relation

$$P_0(x) = 1, \quad P_1(x) = x,$$
$$(n + 1)P_{n+1}(x) = (2n + 1)xP_n(x) - nP_{n-1}(x) \qquad (5.22)$$

These polynomials are *orthogonal* because

$$\int_{-1}^{1} P_j(x)P_k(x)dx = 0$$

whenever $j \neq k$. (This is the generalization to functions of the idea of two vectors being orthogonal, or perpendicular.)

The first 4 Legendre polynomials are

$$P_0(x) = 1, \quad P_1(x) = x$$
$$P_2(x) = \frac{1}{2}\left(3x^2 - 1\right)$$
$$P_3(x) = \frac{1}{2}\left(5x^3 - 3x\right)$$
$$P_4(x) = \frac{1}{8}\left(35x^4 - 30x^2 + 3\right)$$

Exercises:
Section 5.2

1 Derive the formula for the basic trapezoid rule by finding coefficients so that

$$\int_a^b f(x)dx = \alpha f(a) + \beta f(b)$$

for $f(x) = 1, x$.

2 Find the interpolatory quadrature rule for the interval $[-2, 2]$ using the nodes $-1, 0, +1$. What is its degree of precision?

3 Repeat Exercise 2 for the nodes $-2, -1, 0, +1, +2$. Use the resulting formula to estimate the area of a semicircle of radius 2.

4 Use the midpoint, trapezoid and Simpson's rules to estimate $\displaystyle\int_0^1 \frac{1}{1+x^2}\,dx$. Compare the results to the true value $\pi/4$.

5 Use the midpoint, trapezoid and Simpson's rules to estimate $\displaystyle\int_1^2 \frac{1}{x}\,dx$. Compare the results to the true value $\ln 2$.

6 Find the Gaussian quadrature rule using 3 nodes in $[0,1]$. Use this rule to estimate $\displaystyle\int_0^1 \frac{1}{1+x^2}\,dx$ and $\displaystyle\int_0^1 \frac{1}{x+1}\,dx$. Compare the results with those obtained in Exercises 4 and 5.

7 Use (5.15) to verify that the first 4 Legendre polynomials are given by

$$P_0(x) = 1, \quad P_1(x) = x, \quad P_2(x) = \frac{1}{2}\left(3x^2 - 1\right)$$

$$P_3(x) = \frac{1}{2}\left(5x^3 - 3x\right), \quad P_4(x) = \frac{1}{8}\left(35x^4 - 30x^2 + 3\right)$$

8 Find the roots of the Legendre polynomial P_4 and hence find the Gaussian quadrature rule on $[-1,1]$ using 4 nodes. Use it to estimate $\displaystyle\int_{-1}^1 \frac{1}{1+x^2}\,dx$.

5.3 Composite formulas

The accuracy obtained from the trapezoid rule (or any other quadrature formula) can be improved by using the fact that

$$\int_a^b f(x)dx = \int_a^c f(x)dx + \int_c^b f(x)dx$$

for some intermediate point $c \in (a,b)$. For example, using $c = 1/4$ for the trapezoid rule in Example 2, we get

$$N_{0,1}(1/2) \approx \frac{1}{2} + \frac{1}{\sqrt{2\pi}}\frac{1/4}{2}\left[\exp(0) + \exp\left(-(1/4)^2/2\right)\right]$$

$$+ \frac{1}{\sqrt{2\pi}}\frac{1/4}{2}\left[\exp\left(-(1/4)^2/2\right) + \exp\left(-(1/2)^2/2\right)\right]$$

$$= 0.690543$$

reducing the error to about 0.0009. That is, by halving the steplength h, we have reduced the error to approximately one-quarter of its previous value. This, too, is consistent with our claimed error formula (5.11). We shall return to this empirical justification for the error formula shortly.

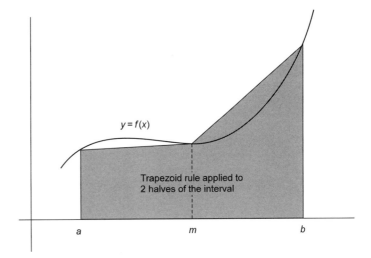

Figure 5.5   Trapezoid rule with 2 intervals

We can rewrite the formula for the trapezoid rule using 2 subdivisions of the original interval (Figure 5.5) to obtain

$$\int_a^b f(x)dx \approx \frac{h}{2}[f(a) + f(a+h)] + \frac{h}{2}[f(a+h) + f(b)] \qquad (5.16)$$

$$= \frac{h}{2}[f(a) + 2f(a+h) + f(b)]$$

where now $h = (b-a)/2$. Geometrically, it is easy to see that this formula is just the average of the left and right sums for this partition of the original interval. Algebraically, this is also easy to see: (5.16) is exactly this average since the corresponding left and right sums are just $h[f(a) + f(a+h)]$ and $h[f(a+h) + f(b)]$.

This provides an easy way to obtain the general formula for the *composite trapezoid rule* T_N using N subdivisions of the original interval. With $h = (b-a)/N$, the left and right sums are given by (5.2) and (5.3), and averaging these we get

$$\int_a^b f(x)dx \approx T_N = \frac{1}{2}(L_N + R_N)$$

$$= \frac{1}{2}\left[h\sum_{k=0}^{N-1} f(a+kh) + h\sum_{k=1}^{N} f(a+kh) \right]$$

$$= \frac{h}{2}\left[f(a) + 2\sum_{k=1}^{N-1} f(a+kh) + f(b) \right] \qquad (5.14)$$

Program   **MATLAB implementation of the composite trapezoid rule**

```
function TN=trapsum(fcn,a,b,N);
% Computes the trapezoid rule for the integral of
% the function fcn over [a,b] using N subdivisions
% The function fcn must be stored as an m-file
h=(b-a)/N;
s=(feval(fcn,a)+feval(fcn,b))/2;
for k=1:N-1
   s=s+feval(fcn,a+k*h);
end
TN=s*h;
```

Here s accumulates the sum of the function values which are finally multiplied by the steplength h to complete the computation. In the version above the summation is done using a loop. The explicit loop can be avoided by using MATLAB's sum function as follows:

```
function TN=trapN(fcn,a,b,N);
% Computes the trapezoid rule for the integral of
% the function fcn over [a,b] using N subdivisions
% The function fcn must be stored as an m-file
h=(b-a)/N;
x=a+(1:N-1)*h;
TN=h*((feval(fcn,a)+feval(fcn,b))/2+sum(feval(fcn,x)));
```

For the latter version it is essential that the function fcn is written for vector inputs, and, if you are using the *Student Edition* of MATLAB, that the number of subdivisions is within the vector size limit. (This limit is $8192 = 2^{13}$ for *Student Edition* Version 4, and $16384 = 2^{14}$ for Version 5.) The first version of the algorithm trapsum.m has no such restriction since it makes no use of MATLAB arrays.

At this stage, we revisit the error formula for the trapezoid rule. First, we observe that the error formula (5.11) remains valid for the composite trapezoid rule – except with $h = (b - a)/N$ of course. The proof of this result is left as an exercise. A similar result for Simpson's rule will be proved later.

Next, we try to justify this error formula empirically. Clearly, the error should depend in some way on the function to be integrated. Since the trapezoid rule integrates polynomials of degree 1 exactly, it is reasonable to suppose that this dependence is on the second derivative. (After all $f''(x) \equiv 0$ for any polynomial of degree 1. Also the error in Lagrange interpolation using a polynomial of degree 1 is dependent on the second derivative.) Furthermore, if a function f varies slowly, we expect $\int_a^b f(x)dx$ to be approximately proportional to the length of the interval

$[a, b]$. Using a constant steplength, we would also expect the trapezoid rule estimate of this integral to be proportional to $b - a$. In that case the error, which is just the difference of these quantities, should be, too. That the error depends on the steplength seems natural and so we may reasonably suppose that the error

$$E(T_N) = \int_a^b f(x)dx - T_N = c(b - a)h^k f''(\xi)$$

for some 'mean value point' ξ, some power k, and some constant c.

By fixing $f(x) = x^2$ so that f'' is constant, and fixing $[a, b]$, we can find suitable values for k and c. It is convenient to choose $[a, b] = [0, 3]$ so that $\int_a^b f(x)dx = 9$. The following MATLAB loop provides a table of values of the errors $E(T_N)$ for this integral with $N = 1, 2, 4, 8$.

```
» N=1;
» for k=1:4
  I=trapN('sqr',0,3,N);
  [N,9-I]
  N=2*N;
end
```

The results, and the ratios of successive errors, are

N	$E(T_N)$	$E(T_N)/E(T_{2N})$
1	-4.5	4.0
2	-1.125	4.0
4	-0.28125	4.0
8	-0.0703125	

Clearly, the error is being reduced by a factor of 4 each time the number of subdivisions is doubled, or, equivalently, h is halved. This provides strong evidence that $k = 2$. Finally, using $N = 1$, we have $E(T_N) = -4.5$ and $b - a = 3, h = 3$, and $f''(\xi) = 2$ so that the constant c must satisfy

$$c(3)(3^2)(2) = -4.5$$

from which we deduce that $c = -1/12$, as stated in (5.11).

Note: This is *not* a proof of this error formula. The arguments presented merely make it reasonable, in at least this case, and provide evidence in support of the constants in the formula. Similar arguments can be given in support of the other error formulas (5.10) and (5.12) for the midpoint and Simpson's rules, but, as was said earlier, the proofs of these results are more complicated than is desirable here.

In using the composite midpoint rule, we use the midpoints of each of the subdivisions. If we denote $a + kh$ in the trapezoid rule by x_k, then the midpoint rule

Figure 5.6   Nodes for composite trapezoid and midpoint rules

using the same N subdivisions of $[a, b]$ uses the nodes $y_k = (x_{k-1} + x_k)/2 = a + \left(k - \frac{1}{2}\right)h$ for $k = 1, 2, \ldots, N$. Thus, we obtain

$$\int_a^b f(x)dx \approx M_N = h\sum_{k=1}^N f(y_k) = h\sum_{k=1}^N f\left(a + \left(k - \frac{1}{2}\right)h\right) \tag{5.18}$$

The distribution of these nodes and those for the trapezoid rule is illustrated in Figure 5.6.

The composite version of Simpson's rule using $h = (b - a)/2N$ uses all the points for the composite midpoint rule (5.15) as well as those of the composite trapezoid rule (5.14). Rewriting these two rules with this value of h, we get

$$M_N = 2h\sum_{k=1}^N f(y_k) = h\sum_{k=1}^N f(a + (2k - 1)h)$$

$$T_N = h\left[f(a) + 2\sum_{k=1}^{N-1} f(x_k) + f(b)\right] = 2h\left[f(a) + 2\sum_{k=1}^{N-1} f(a + 2kh) + f(b)\right]$$

Now applying Simpson's rule to each of the intervals $[x_{k-1}, x_k]$ we get the composite Simpson's rule formula

$$S_{2N} = \frac{h}{3}\left[\begin{array}{l} f(a) + 4f(a + h) + 2f(a + 2h) + \cdots \\ +2f(a + 2(N - 1)h) + 4f(a + (2N - 1)h) + f(b) \end{array}\right]$$

$$= \frac{h}{3}\left[f(a) + 4\sum_{k=1}^N f(y_k) + 2\sum_{k=1}^{N-1} f(x_k) + f(b)\right] \tag{5.19}$$

$$= \frac{T_N + 2M_N}{3} \tag{5.20}$$

The final version (5.20) of this formula can be used to create efficient implementations of Simpson's rule. We shall discuss this further in the next section. The first form (5.19) is very easy to implement in MATLAB.

Note that Simpson's rule is defined only for an even number of subdivisions, since each subinterval on which Simpson's rule is applied must itself be subdivided into two pieces.

The following code implements this version using explicit loops. As with the trapezoid rule, the sum function can be used to shorten the code.

| **Program** | **MATLAB code for composite Simpson's rule using a fixed number of subdivisions** |

```
function SN=Simpsum(fcn,a,b,N);
% Computes the Simpson rule for the integral of
% the function fcn over [a,b] using N subdivisions
% N MUST be even
% The function fcn must be stored as an m-file
h=(b-a)/N;
s=feval(fcn,a)+feval(fcn,b);
for k=1:2:N-1
  s=s+4*feval(fcn,a+k*h);
end
for k=2:2:N-2
  s=s+2*feval(fcn,a+k*h);
end
SN=s*h/3;
```

It would be easy to reinitialize the sum so that only a single loop is used, but the structure is kept simple here to reflect (5.19) as closely as possible.

Example 6   Use the midpoint, trapezoid and Simpson's rules to estimate

$$I = \int_0^1 \frac{1}{1+x^2}\,dx$$

using $N = 1, 2, 4, 8, 16$ subdivisions. The true value is $I = \pi/4 \approx 0.7853982$. Compare the errors in the approximations

Using the trapsum and Simpsum m-files, and a similar program for the midpoint rule, we obtain the following table of results:

N	T_N	M_N	S_N	$E(T_N)$	$E(M_N)$	$E(S_N)$
1	0.7500	0.8000		0.0354	−0.0146	
2	0.7750	0.7906	0.783333	0.0104	−0.0052	0.0021
4	0.7828	0.7867	0.785392	0.0026	−0.0013	6×10^{-6}
8	0.7847	0.7857	0.785398	0.0007	−0.0003	4×10^{-8}
16	0.7852	0.7855	0.785398	0.0002	−0.0001	6×10^{-10}

It is immediately apparent in Example 6 that the Simpson's rule estimates are *much* more accurate for the same numbers of subdivisions – and that their errors are being reduced faster, so that the advantage is increasing at each stage. It is for this reason that we shall concentrate on this method as the basis for our practical numerical integration techniques in the next section.

First we establish the following important theorem on the errors in composite Simpson's rule estimates. This result will also be very helpful in our practical algorithm. This theorem establishes that the error formula (5.12) for the basic Simpson's rule remains valid for the composite rule.

Theorem 2 *Suppose the integral $I = \int_a^b f(x)dx$ is estimated by Simpson's rule S_N using N subdivisions of $[a, b]$ and suppose that $f^{(4)}$ is continuous; then the error in this approximation is given by*

$$I - S_N = -\frac{(b-a)h^4}{180}f^{(4)}(\xi) \qquad (5.21)$$

for some $\xi \in (a, b)$ and where $h = (b - a)/N$

Proof First we note that N is necessarily even, and we denote $N/2$ by M. The composite Simpson rule is equivalent to applying the basic Simpson's rule to each of the M intervals $[x_k, x_{k+1}]$ where $x_k = a + 2kh$ ($k = 0, 1, \ldots, M - 1$). By (5.12)

$$\int_{x_k}^{x_{k+1}} f(x)dx - \frac{h}{3}[f(x_k) + 4f(x_k + h) + f(x_{k+1})] = -\frac{(x_{k+1} - x_k)h^4}{180}f^{(4)}(\xi_k)$$

for some $\xi_k \in (x_k, x_{k+1})$.

Since $x_{k+1} - x_k = 2h$, it follows that the total error is given by

$$E(S_N) = -\frac{2h^5}{180}\sum_{k=0}^{M-1}f^{(4)}(\xi_k)$$

Because this fourth derivative is continuous, and therefore bounded, it follows from the intermediate value theorem that there exists a point $\xi \in (a, b)$ such that

$$Mf^{(4)}(\xi) = \sum_{k=0}^{M-1}f^{(4)}(\xi_k)$$

and so we may write

$$E(S_N) = -\frac{2Mh^5}{180}f^{(4)}(\xi)$$

Finally, we note that $2Mh = Nh = (b - a)$ so that

$$E(S_N) = -\frac{(b-a)h^4}{180}f^{(4)}(\xi)$$

as required. ∎

Example 7 Obtain an error bound for the Simpson's rule estimates found in Example 6

It is easy but messy to show that for $f(x) = 1/(1+x^2)$ on $[0,1]$, the fourth derivative is bounded by 24. That is, for $x \in [0,1]$, we have

$$\left| f^{(4)}(x) \right| \leq 24$$

In each case, $b - a = 1$ and $h^4 = 1/N^4$. Therefore, with $N = 2, 4, 8, 16$, we obtain error bounds

$$\frac{24}{180N^4} = \left(\frac{24}{180(2^4)}, \frac{24}{180(4^4)}, \frac{24}{180(8^4)}, \frac{24}{180(16^4)} \right)$$

$$\approx \left(8 \times 10^{-3}, 5 \times 10^{-4}, 3 \times 10^{-5}, 2 \times 10^{-6} \right)$$

Note that these error bounds are all significantly larger than the actual errors – but, of course, we know that only because we know the exact answer in this case. The important point is that these bounds are provable – independent of the true answer. It is this that allows Theorem 2 to be used as the basis of practical numerical integration routines.

Exercises: Section 5.3

1 Write a MATLAB m-file for the midpoint rule using N subdivisions for $\int_a^b f(x)\,dx$. Use it to find estimates of $\ln 3 = \int_1^3 (1/x)\,dx$ using $N = 1, 2, 4, 8, 16$.

2 Use your answers for Exercise 1 to justify the claim that the error in the midpoint rule is proportional to h^2.

3 Derive the error formula for the composite trapezoid rule. Use this formula to obtain bounds on the trapezoid rule errors in Example 5.

4 Repeat Exercises 1 and 2 for the trapezoid rule.

5 Repeat the computation of Exercise 1 for Simpson's rule with $N = 2, 4, 8, 16$. Use the results to check that the error is proportional to h^4.

6 Use the midpoint, trapezoid and Simpson's rules to estimate

$$N_{0,1}(2) = \frac{1}{2} + \frac{1}{\sqrt{2\pi}} \int_0^2 \exp\left(-x^2/2 \right) dx$$

using $N = 10$ subdivisions.

7 Repeat Exercise 7 using $N = 20, 40, 80, 160$. Compare the results with the true answer 0.97724987 to 8 decimals. (This is available in MATLAB as (1+erf(sqrt(2)))/2.) Verify that the errors are behaving as expected.

5.4 Practical numerical integration

In this section, we are primarily concerned with practical algorithms for automatic numerical integration, where our objective is to compute

$$I = \int_a^b f(x)dx$$

with an error smaller than some prescribed tolerance ε. The methods we discuss will all be based on using elementary rules such as the trapezoid and Simpson's rules, mostly the latter.

In Section 5.3, we saw that the error in the composite Simpson's rule estimate of I is given by (5.18)

$$I - S_N = -\frac{(b-a)h^4}{180}f^{(4)}(\xi)$$

where ξ is some (unknown) mean value point.

This gives us our first approach to practical numerical integration which uses the following basic steps:

1   Find a bound M_4 for $\left| f^{(4)}(x) \right|$ on $[a, b]$.
2   Choose N such that

    (a) N is even, and
    (b)   with $h = (b-a)/N$, we have

$$\left| \frac{(b-a)h^4}{180}M_4 \right| < \varepsilon \tag{5.22}$$

3   Evaluate S_N.

The first step is (typically) a paper and pencil exercise but the rest is easily implemented as an automatic process using the Simpson's rule function that we already have. The first part can be achieved using a computer algebra system such as Maple,[1] or the Symbolic Toolbox that is a part of the *Student Edition* of MATLAB.

In applying (5.22), it is convenient to replace h by $(b-a)/N$, so that we require

$$\left| \frac{(b-a)^5}{180N^4}M_4 \right| < \varepsilon$$

from which we deduce the requirement

$$N > |b-a|\sqrt[4]{\frac{|b-a|M_4}{180\varepsilon}} \tag{5.23}$$

We illustrate the process with an example.

---

[1] Maple is a registered trademark of Waterloo Maple, Inc.

Example 8 Use the composite Simpson's rule to evaluate

$$I = \int_0^1 \frac{1}{1+x^2}\,dx$$

with an error smaller than 10^{-8}

We saw in Example 7 in the previous section that $\left|f^{(4)}(x)\right| \leq 24$ on $[0,1]$ and so we may take $M_4 = 24$. It follows from (5.23) that we desire

$$N > \left[\frac{24}{180 \times 10^{-8}}\right]^{1/4} = 100\sqrt[4]{2/15} \approx 60.4275$$

N must be an even integer, and so we choose $N = 62$.
The following MATLAB command yields the result shown.

```
» S=simpsum('runge',0,1,62)
S = 0.785398163397273.
```

The true value is $\pi/4 = 0.785398163397448$ so that the actual error is only about 2×10^{-13} which is much smaller than the required tolerance.

It is common that the actual error will be much smaller than the predicted error. The main cause for this is that the bound on the fourth derivative may be much greater than the value through most of the range of integration, and, in particular, M_4 may be much greater than the value of $\left|f^{(4)}(\xi)\right|$ in (5.21). Nonetheless, this process gives us a reliable and fairly inexpensive method for computing integrals such as this one.

We can automate all of this except for the evaluation of M_4. The following m-file implements all of this procedure once a value of M_4 is known. It simply computes N using (5.23) and then calls the previously defined simpsum function.

Program | MATLAB m-file for computing $\int_a^b f(x)dx$ with error less than **eps** using Simpson's rule

```
function I=simpson(fcn,a,b,M4,eps)
% M4 is a bound for the fourth derivative of fcn on [a,b]
% eps is the required accuracy
L=abs(b-a); N=ceil(L*sqrt(sqrt(L*M4/180/eps)));
% N must be even
if mod(N,2)==1; N=N+1; end
I=simpsum(fcn,a,b,N);
```

Note the use of abs(b-a). This allows the possibility that $b < a$ which is mathematically well defined, and which is also handled perfectly well by Simpson's rule. If $b < a$, then the steplength h will be negative and the appropriately signed result will be obtained.

The use of this m-file for the integral of Example 7 is:

```
» S=simpson('runge',0,1,24,1e-8);
```

which gives the same result as before.

In the example above it was reasonably easy to obtain the fourth derivative, and then a bound for this function on $[a, b]$. This will not always be so easy. In such cases there is a simple alternative to the procedure described above. We can start with a small number of subdivisions of the range of integration and continually double this until we get two results which agree to within our tolerance. The error formula (5.21) gives us good reason to believe that an answer obtained in this way will indeed satisfy our accuracy requirement.

Consider (5.21) for Simpson's rule using first N and then $2N$ intervals. Writing $h = (b - a)/2N$, we get

$$E(S_N) = -\frac{(b - a)(2h)^4}{180} f^{(4)}(\xi_1) \qquad (5.24)$$

$$E(S_{2N}) = -\frac{(b - a)h^4}{180} f^{(4)}(\xi_2) \approx \frac{E(S_N)}{16} \qquad (5.25)$$

if $f^{(4)}$ is approximately constant over $[a, b]$. If $|S_N - S_{2N}| < \varepsilon$, and $16(I - S_{2N}) \approx (I - S_N)$, then it follows that

$$|I - S_{2N}| \approx \varepsilon/15 \qquad (5.26)$$

so that we expect the second of these estimates to be well within the tolerance.

Example 9 **Recompute the integral** $I = \int_0^1 \frac{1}{1+x^2} dx$ **starting with** $N = 2$ **and repeatedly doubling** N **until two Simpson's rule estimates agree to within** 10^{-10}

The while loop below produces the results shown:

```
» N=2;
» newS=simpsum('runge',0,1,N);
» oldS=0;
» while abs(oldS-newS)>1e-10
   oldS=newS; N=2*N;
   newS=simpsum('runge',0,1,N);
  [N,newS]
 end
```

4	0.785392156862745
8	0.785398125614677
16	0.785398162806206
32	0.785398163388209
64	0.785398163397304

which is essentially the same answer as we obtained previously.

Note that the computation of Example 9 required about twice as many points in total, since the earlier function evaluations are repeated. It is possible to remove this repetition by making efficient use of the midpoint and trapezoid rules to construct Simpson's rule as in (5.20) so that only the new nodes are evaluated at each iteration. We shall not concern ourselves with this aspect here.

It is easy to construct an automatic integration m-file based on this repeated doubling of the number of intervals. That is left as an exercise.

The comparison of the errors in two successive Simpson's rule estimates, (5.24) and (5.25), of an integral can be utilized further to obtain greater accuracy at virtually no greater computational cost. The approximate equation

$$16(I - S_{2N}) \approx (I - S_N)$$

can be rearranged to yield

$$I \approx \frac{16 S_{2N} - S_N}{15} \tag{5.27}$$

and this process should eliminate the most significant contribution to the error.

Using $N = 2$ in (5.27) and with $h = (b - a)/2N$, we get

$$15I \approx \frac{16h}{3}\{f(a) + 4f(a + h) + 2f(a + 2h) + 4f(a + 3h) + f(b)\}$$
$$- \frac{2h}{3}\{f(a) + 4f(a + 2h) + f(b)\}$$

which simplifies to yield

$$I \approx \frac{2h}{45}[7f(a) + 32f(a + h) + 12f(a + 2h) + 32f(a + 3h) + 7f(b)] \tag{5.28}$$

This formula is in fact the interpolatory quadrature rule using five equally spaced nodes. It could therefore have been derived in the same manner as the other interpolatory rules. The error in using (5.28) is of the form $C(b - a)h^6 f^{(6)}(\xi)$. The above process of halving the steplength and then eliminating the most significant error contribution could be repeated using (5.28) as the basic rule. We would obtain a still more accurate quadrature formula. This principle is the basis of *Romberg integration*.

The process of Romberg integration can be (and usually is) started from the trapezoid rule rather than Simpson's rule. In that case we would simply rediscover Simpson's rule as the result of eliminating the second-order error term between T_1 and T_2. Specifically, we find

$$S_{2N} = \frac{4T_{2N} - T_N}{3} \tag{5.29}$$

(The verification of this result is left as an exercise.)

In order to describe Romberg integration more systematically, it is convenient to introduce a notation for the various estimates obtained. These will be arranged as elements of a matrix (actually a lower triangular matrix) as follows. The first column has entries $R_{n,1}$ which are the trapezoid rule estimates using 2^{n-1} subdivisions:

$$R_{n,1} = T_{2^{n-1}}$$

The second column has the results that would be obtained using Simpson's rule, so that, for $n \geq 2$, we have

$$R_{n,2} = S_{2^{n-1}} = \frac{4T_{2^{n-1}} - T_{2^{n-2}}}{3} = \frac{4R_{n,1} - R_{n-1,1}}{3}$$

Now applying the improved estimate (5.27), or equivalently, (5.28), we get

$$R_{n,3} = \frac{16R_{n,2} - R_{n-1,2}}{15}$$

for $n \geq 3$.

Continuing in this way we obtain the array

$$
\begin{array}{cccccc}
R_{1,1} \\
R_{2,1} & R_{2,2} \\
R_{3,1} & R_{3,2} & R_{3,3} \\
R_{4,1} & R_{4,2} & R_{4,3} & R_{4,4} \\
R_{5,1} & R_{5,2} & R_{5,3} & R_{5,4} & R_{5,5} \\
\vdots & \vdots & \vdots & \vdots & \vdots & \ddots \\
\vdots & \vdots & \vdots & \vdots & \vdots & \vdots & \ddots
\end{array}
$$

The general formula for generating entries further to the right in this table is obtained by extending the arguments presented above. It is

$$R_{n,k+1} = \frac{4^k R_{n,k} - R_{n-1,k}}{4^k - 1} \tag{5.30}$$

The following MATLAB code illustrates the implementation of this process to compute the array as far as $R_{M,M}$.

Program | **MATLAB m-file for Romberg integration as far as the entry $R_{M,M}$ where M is some maximum level to be used**

```
function R=romberg1(fcn,a,b,ML)
% Romberg integration based on the trapezoid rule
% ML is the maximum level of the array
% Output here is the full Romberg array
N=1;
R(1,1)=(b-a)*(feval(fcn,a)+feval(fcn,b))/2;
for L=2:ML
  M=midptsum(fcn,a,b,N);
  R(L,1)=(R(L-1,1)+M)/2;
  for k=1:L-1
  R(L,k+1)=(4^k*R(L,k)-R(L-1,k))/(4^ k-1);
  end
  N=N*2;
end
```

Note the use of the appropriate midpoint rule to update the trapezoid rule R(L,1). It is reasonably easy to verify that

$$T_{2N} = \frac{T_N + M_N}{2}$$

and this permits a more efficient computation of subsequent trapezoid rule estimates by eliminating the need to re-evaluate the function at nodes that have already been used. (See Exercise 3 for a verification of this identity.)

Example 10   Use Romberg integration with 6 levels to estimate the integrals

(a) $\displaystyle\int_0^1 \frac{1}{1+x^2}\,dx$, and

(b) $\displaystyle\frac{1}{\sqrt{2\pi}}\int_0^2 \exp\left(-x^2/2\right)dx$

The following MATLAB commands can be used with the m-file romberg1 above to yield the results shown.

(a) » R=romberg1('runge',0,1,6)

R =

0.75000	0	0	0	0	0
0.77500	0.78333	0	0	0	0
0.78279	0.78539	0.78553	0	0	0
0.78475	0.78540	0.78540	0.78540	0	0
0.78524	0.78540	0.78540	0.78540	0.78540	0
0.78536	0.78540	0.78540	0.78540	0.78540	0.78540

and

(b) » R=romberg1('intex',0,2,6)

R =

0.45293	0	0	0	0	0
0.46844	0.47361	0	0	0	0
0.47501	0.47720	0.47744	0	0	0
0.47669	0.47725	0.47725	0.47725	0	0
0.47711	0.47725	0.47725	0.47725	0.47725	0
0.47721	0.47725	0.47725	0.47725	0.47725	0.47725

The zeros represent the fact that the MATLAB code uses a matrix to store R and these entries are never assigned values. In the mathematical algorithm these entries simply do not exist.

The usual way in which Romberg integration is used is not to perform a fixed number of steps but rather to use it iteratively until two successive values agree to within a specified tolerance. Usually this agreement is sought among entries on the diagonal of the array, or between the values $R_{n,n-1}$ and $R_{n,n}$. The program above is easily modified for this purpose. This modification is explored in the exercises.

We conclude this section with a brief look at *adaptive quadrature*. Adaptive integration algorithms use the basic principle of subdividing the range of integration but without the earlier insistence on using uniform subdivisions throughout the range. To simplify the discussion, we shall restrict our attention to the adaptive use of Simpson's rule.

There are two fundamentally different approaches available to us.

1 Decide on an initial subdivision of the interval into N subintervals $[x_k, x_{k+1}]$ with

$$a = x_0 < x_1 < x_2 < \cdots < x_N = b$$

Use the composite Simpson's rule with continual doubling of the number of intervals in each of the N original subintervals in turn. (Typical choices of the initial subdivision are 5 or 20 equal subintervals.)

2 Start with the full interval. Apply Simpson's rule to the current interval and to each of its two halves. If the results agree to within the required accuracy, then

we accept this partial result, and move on to the next subinterval. If the estimates do not agree to the required accuracy, then continue working with one half, leaving the other half to be dealt with later.

The details of these methods can vary immensely. Also it should be apparent that any basic rule, or indeed Romberg integration, could be used with these same principles. We shall illustrate both approaches with a simple example.

Example 11 **Compute**

$$\pi = \int_0^1 \frac{4}{1+x^2}\,dx$$

with error less than 10^{-6} using adaptive Simpson's rule with $N = 5$ initial subdivisions

On each of the intervals $[0, 1/5]$, $[1/5, 2/5]$, $[2/5, 3/5]$, $[3/5, 4/5]$ and $[4/5, 1]$ we use continual halving of the steplengths until the Simpson's rule estimates agree to within $10^{-6}/5$.

On $[0, 1/5]$ we get the results:

$$N = 2 \qquad S_2 = 0.78959126681899$$
$$N = 4 \qquad S_4 = 0.789582790606273$$
$$N = 8 \qquad S_8 = 0.789582273655695$$
$$N = 16 \quad S_{16} = 0.789582241537532$$

On the remaining intervals, the converged values are:

$$[1/5, 2/5] \quad S_8 = 0.732443273491736$$
$$[2/5, 3/5] \quad S_{16} = 0.63965249167752$$
$$[3/5, 4/5] \quad S_{16} = 0.537285766868543$$
$$[4/5, 1] \quad S_8 = 0.44262887640553$$

and summing these, we obtain the approximation $\pi \approx 3.14159264998086$ which has an error of about 3.6×10^{-8}.

If there were no repetitions of any function evaluations, this would have used $17 + 8 + 16 + 16 + 8 = 65$ points in all. In this particular case, Romberg integration using the same total number of points yields $\pi \approx 3.14159265358972$, which has an error of only about 7×10^{-14}.

The actual error is much smaller than the required tolerance. One reason for this is that, as we saw in our discussion of Romberg integration, if $|S_N - S_{2N}| \approx \varepsilon$, then we expect $|I - S_{2N}| \approx \varepsilon/15$. The local tolerance requirement in Example 11 is therefore much smaller than is needed for the overall accuracy.

Example 12   Use adaptive Simpson's rule quadrature with the interval being continually split to evaluate the integral of Example 10

We begin with the whole interval $[0, 1]$. For any interval for which we do not have the required agreement between S_2 and S_4, we work on just the left-hand half (for which we already have S_2 as part of the previous S_4) until agreement is reached. We then proceed to the next piece.

We first get agreement on $[0, 1/32]$ with $S_4 = 0.124959334$. The right-hand half of the most recently divided interval is then $[1/32, 1/16]$ on which we immediately get agreement. Similarly agreement is immediate on the next piece $[1/16, 1/8]$. Eventually, we obtain

$$\pi \approx 3.1415926$$

having used intervals with end-points 0, 1/32, 1/16, 1/8, 3/16, 1/4, 3/8, 1/2, 9/16, 5/8, 11/16, 3/4, 7/8, and 1. A total of 53 function evaluations is needed – assuming that all those from 'rejected' parts of intervals are retained for later use.

Efficient programming of adaptive quadrature as in Example 12 is more complicated than the other schemes discussed here. It has significant advantages when the integrand varies more than in this example, but for relatively well behaved functions such as we have encountered here, the more easily programmed Romberg technique or the use of an initial subdivision, as in Example 11, is probably to be preferred.

**Exercises:
Section 5.4**

1   Show that

$$S_2 = \frac{4T_2 - T_1}{3}$$

and verify the result for $\int_0^1 1/(1 + x^2)\,dx$.

2   Use the result of the previous exercise to prove (5.29):

$$S_{2N} = \frac{4T_{2N} - T_N}{3}$$

3   Show that $T_2 = (T_1 + M_1)/2$ and therefore that

$$T_{2N} = \frac{T_N + M_N}{2}$$

4   Use Romberg integration with four levels to estimate $\displaystyle\int_0^2 \frac{1}{1 + x^2}\,dx$.

5   Write a MATLAB program to perform Romberg integration iteratively until two diagonal entries $R_{n-1,n-1}$ and $R_{n,n}$ agree to within a specified accuracy. Use your program to estimate $\displaystyle\int_{-1}^1 \frac{1}{1 + x^2}\,dx$ with a tolerance of 10^{-8}.

6 Write a MATLAB m-file which applies composite Simpson's rule with repeated doubling of the number of intervals in each of five initial subintervals. Verify the results of Example 9.

7 Modify your MATLAB program for Exercise 6 to perform adaptive integration using the composite Simpson's rule on each interval of a specified uniform partition of the original interval into N pieces. Use it to evaluate the integral of Exercise 5 with $N = 5$, 10 and 20 initial subintervals.

5.5 Improper integrals

In this section, we consider the numerical evaluation of improper integrals. We shall concentrate almost entirely on integrals with an infinite range of integration, and we specialize this to the situation of computing

$$I(a, \infty) = \int_a^\infty f(x)dx \tag{5.31}$$

There are advanced methods which are specifically developed for this situation, but we shall consider how to take advantage of the methods already developed to evaluate such integrals. We shall not concern ourselves directly with establishing the convergence of the integrals under consideration, although the methods we adopt will frequently use the proof of convergence in order to 'bound the tail' of the integral.

We shall assume throughout this section that the integral in question is convergent. Our objective is to compute such an integral with an overall error less than some tolerance ε. The basic idea is to separate the integral into 2 pieces:

$$I = \int_a^b f(x)dx + \int_b^\infty f(x)dx$$

Because the integral converges, it follows that the second piece

$$I(b, \infty) = \int_b^\infty f(x)dx \to 0 \text{ as } b \to \infty$$

and therefore that we can find a value of b for which

$$\int_b^\infty f(x)dx < \frac{\varepsilon}{2} \tag{5.32}$$

The basic idea is that we choose a number b such that the 'tail' of the integral is small. The situation is illustrated in Figure 5.7.

We want the shaded area to be small enough to neglect yet still compute the integral to the desired accuracy. Suppose we wish to find $I(a, \infty)$ in (5.31) with error no greater than ε. Then, using (5.32)

$$\left| I(a, \infty) - \int_a^b f(x)dx \right| = I(b, \infty) < \frac{\varepsilon}{2}$$

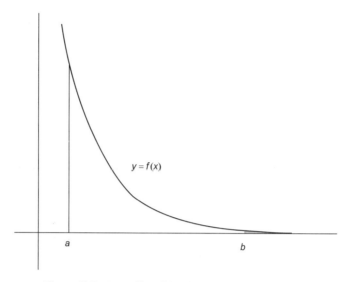

Figure 5.7 Bounding the tail of an infinite integral

If we now estimate $I(a,b) = \int_a^b f(x)dx$ with error smaller than $\varepsilon/2$, it will follow that this computed value, S, approximates $I(a,\infty)$ with error less than ε, since

$$|I(a,\infty) - S| = \left| \int_a^\infty f(x)dx - S \right| = \left| \int_a^b f(x)dx - S + \int_b^\infty f(x)dx \right|$$

$$\leq \left| \int_a^b f(x)dx - S \right| + I(b,\infty)$$

$$< \frac{\varepsilon}{2} + \frac{\varepsilon}{2} = \varepsilon$$

Example 13 Compute the infinite integral $\displaystyle\int_1^\infty \frac{1}{x}e^{-x}dx$ **with an error bounded by** $\varepsilon = 10^{-6}$

That this integral exists is an easy application of the Comparison Test. (Exercise)

First we want to find a number b such that

$$\int_b^\infty \frac{1}{x}e^{-x}dx < \frac{\varepsilon}{2}$$

Now, if $x \geq b > 1$, then

$$\frac{1}{x}e^{-x} < e^{-x}$$

and hence we deduce that

$$\int_b^\infty \frac{1}{x} e^{-x} dx < \int_b^\infty e^{-x} dx$$

$$= \lim_{t \to \infty} \int_b^t e^{-x} dx = \lim_{t \to \infty} e^{-b} - e^{-t}$$

$$= e^{-b}$$

It follows that $\int_b^\infty \frac{1}{x} e^{-x} dx < \frac{\varepsilon}{2}$ if we can find b such that $e^{-b} < \frac{\varepsilon}{2}$.

Taking logarithms, for $\varepsilon = 10^{-6}$, this yields the requirement

$$b > \ln 2 + 6 \ln 10 \simeq 14.5$$

We may safely take $b = 15$.

The next step is to compute

$$\int_1^{15} \frac{1}{x} e^{-x} dx$$

with an error smaller than $\varepsilon/2$. Note that this is an integral that we can not compute exactly by Calculus techniques. However, we can obtain its fourth derivative and then use the error formula for Simpson's rule to guarantee the necessary accuracy in the final answer:

$$\frac{d^4}{dx^4} \frac{1}{x} e^{-x} = \frac{24}{x^5} e^{-x} + \frac{24}{x^4} e^{-x} + \frac{12}{x^3} e^{-x} + \frac{4}{x^2} e^{-x} + \frac{1}{x} e^{-x}$$

is a decreasing function (for $x > 0$). Therefore we may use $x = 1$ to get $M_4 = \dfrac{65}{e} = 23.912$. For simplicity, we shall take $M_4 = 24$ which with $|b - a| = 15 - 1 = 14$ gives

$$N \geq 14 \left(\frac{14 \times 24}{180 \times 10^{-6}/2} \right)^{1/4} \simeq 615.4$$

Therefore $N = 616$ intervals will suffice. (Of course this last calculation could have been part of the MATLAB procedure.) We can now use our Simpson's rule m-file MATLAB to get the final result:

```
» Int=simpsum('imprex',1,15,616)
Int =
    0.21938392392287
```

where the function imprex is defined in the m-file

```
function y=imprex(x)
y=1./x./exp(x);
```

The b value used in this example is not optimal. For $x \geq b > 1$, we used the inequality $\frac{1}{x}e^{-x} < e^{-x}$. However, if we use the stronger inequality

$$\frac{1}{x}e^{-x} < \frac{1}{b}e^{-x}$$

the requirement for b becomes $\frac{1}{b}e^{-b} < \varepsilon/2$. For $\varepsilon = 10^{-6}$, we could then use $b = 12$. The resulting interval will be smaller and consequently the number of steps needed will also be reduced: $N = 456$ would now suffice. For this example, this saving is unimportant, but the same principles apply more widely and can result in significant improvements.

Note: There is nothing particularly special about dividing the tolerance into 2 *equal* pieces. Provided the two tolerances used (for the tail and for Simpson's rule) add up to the desired overall accuracy, we can distribute them as we like. Sometimes there may be some benefit to distributing them differently.

In this text, we do not pursue the related question of handling singularities at one end of the range of integration. Such difficulties can arise when the integrand tends to infinity at one end. For example, the integral $\int_0^1 \frac{1}{\sqrt{x}}dx$ converges but $\frac{1}{\sqrt{x}} \to \infty$ as $x \to 0+$. We could not apply Simpson's rule directly to such an integral because of this. Similar techniques to those outlined for infinite integrals can be used to handle singularities of this nature, but there are additional complications caused by the fact that the derivative used in applying the error bound is also unbounded. This can lead to prohibitively high values for the number of steps. We do not pursue this here.

Exercises: Section 5.5

1 Show that the integral $\int_1^\infty \frac{1}{x}e^{-x}dx$ in Example 13 converges. Use the technique of Example 13 to evaluate it with error less than 10^{-8}.

2 Compute $\int_1^\infty \frac{1}{\sqrt{x}}e^{-x}dx$ with error less than 10^{-6}.

3 Define

$$I(p) = \int_1^\infty \frac{1}{x^p}e^{-x}dx$$

Show that these integrals converge for all $p > 0$. Obtain a table of values of $I(p)$ for $p = \frac{1}{2}, 1, \frac{3}{2}, 2, \ldots, 6$ each accurate within 10^{-8}.

4 Repeat Exercise 3, varying the proportion of the error allocated to the tail and to the finite integral. What differences do you observe in efficiency? Which is most accurate?

5.6 Numerical differentiation

In some respects, the methods of numerical differentiation are similar to those of numerical integration, in that they are typically based on using (in this case, differentiating) an interpolation polynomial. There is, however, one major and important difference. We saw in previous sections that integration is numerically a highly satisfactory operation, with results of high accuracy being obtainable in economical ways. This is a consequence of the fact that integration tends to smooth out the errors of the polynomial approximations to the integrand. Unfortunately this is certainly not the case for differentiation which tends to exaggerate the error in the original polynomial approximation. Figure 5.8 illustrates this clearly.

It is worth noting that although numerical differentiation is likely to be unreliable (especially if high accuracy is needed), its symbolic counterpart is generally fairly easy. Computer Algebra Systems, such as Maple, need program only a small number of basic rules in order to differentiate a wide variety of functions. This is in direct contrast to the situation for integration. We have seen that numerical integration is reasonably easy, but symbolic integration is very difficult to program well. This is partly due to the fact that many elementary functions have no antiderivative that can be written in terms of the same set of elementary functions. Even where this is possible, there are no simple algorithms to determine which integration techniques should be applied. This contrast illustrates at a simple level the need for mathematicians, scientists and engineers to develop a familiarity with both numerical and symbolic computation. The blending of the two in solving more difficult problems remains a very active field of scientific research which is blurring the lines among science, computer science and pure and applied mathematics.

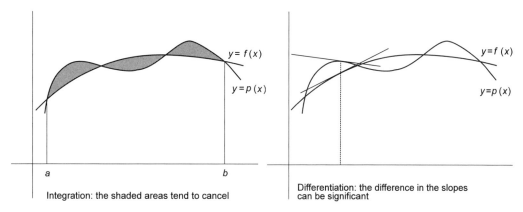

Integration: the shaded areas tend to cancel

Differentiation: the difference in the slopes can be significant

Figure 5.8 Numerical integration and differentiation

For the reasons cited numerical differentiation is best avoided whenever possible. However, we have already seen, in discussing Newton's method and the secant method, that if the true value of the derivative is of secondary importance, then a simple divided difference approximation to the derivative can be satisfactory. We begin our very brief look at numerical differentiation with this approximation.

By definition

$$f'(x_0) = \lim_{h \to 0} \frac{f(x_0 + h) - f(x)}{h}$$

from which we obtain the simple divided difference approximation

$$f'(x_0) \approx \frac{f(x_0 + h) - f(x)}{h} = f[x_0, x_0 + h] \tag{5.33}$$

for some small h. (This, of course, is precisely the approximation used for the secant method, as we saw in Chapter 2.)

Now, by Taylor's theorem (provided that f'' exists)

$$f(x_0 + h) = f(x_0) + hf'(x_0) + \frac{h^2}{2}f''(\theta)$$

for some θ between x_0 and $x_0 + h$. We can now deduce that

$$f'(x_0) - f[x_0, x_0 + h] = f'(x_0) - \frac{f(x_0 + h) - f(x)}{h} = -\frac{h}{2}f''(\theta) \tag{5.34}$$

Therefore the (truncation) error in this approximation is roughly proportional to the steplength h used in its computation. The situation is made worse by the fact that the roundoff error in computing the approximate derivative (5.33) is (roughly) proportional to $1/h$. The overall error therefore is of the form

$$E = ch + \frac{\delta}{h}$$

where c and δ are constants. This error function has the generic shape illustrated in Figure 5.9. This places severe restriction on the accuracy that can be achieved with this formula.

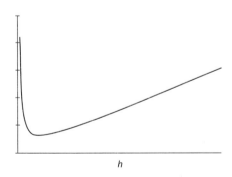

h

Figure 5.9   Error in numerical differentiation

Example 14   Use (5.33) with $h = 0.1, 0.01, \ldots, 10^{-8}$ to estimate $f'(0.7)$ for $f(x) = \ln x$

This is easily achieved with the MATLAB commands:

```
» x0=0.7;
» h=10.^-(1:10);
» df=(log(x0+h)-log(x0))./h;
```

The results produced are:

k	$f\left[x_0, x_0 + 10^{-k}\right]$
1	**1.3**3531392624523
2	**1.41**846349919564
3	**1.42**755199118544
4	**1.428**46939747199
5	**1.42856**122458124
6	**1.42857**040819067
7	**1.42857132**567897
8	**1.42857142**781949
9	**1.42857**137230834
10	**1.42857**170537525

The true value is of course $1/0.7 = 1.42857142857143$. The figures in **bold** type are the correct digits in each approximation. We see exactly the behaviour predicted by Figure 5.9, the error gets steadily smaller for the first several entries and then begins to rise as h continues to decrease. Furthermore, we see the error is initially proportional to the steplength since the number of correct figures increases by one with each iteration.

We also see in Example 14 that we cannot get very high precision using this formula. Similar results would be obtained using negative steps $-h$. (See Exercise 2.)

Improved results can be obtained by using (5.61) with *both* h and $-h$ as follows:

$$f'(x_0) \approx \frac{f(x_0 + h) - f(x)}{h} = f[x_0, x_0 + h]$$

$$f'(x_0) \approx \frac{f(x_0 - h) - f(x)}{-h} = f[x_0, x_0 - h]$$

Averaging these approximations we get

$$f'(x_0) \approx \frac{1}{2}\left[\frac{f(x_0 + h) - f(x)}{h} + \frac{f(x_0 - h) - f(x)}{-h}\right]$$

$$= \frac{f(x_0 + h) - f(x_0 - h)}{2h} = f[x_0 - h, x_0 + h] \tag{5.35}$$

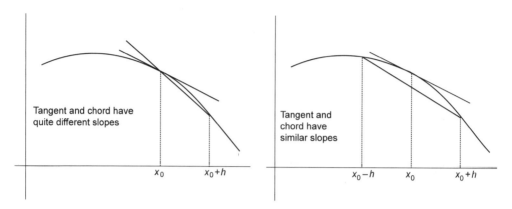

Figure 5.10   Contrast between (5.33) and (5.35)

Assuming that $f^{(3)}$ is continuous in the interval $[x_0 - h, x_0 + h]$, we can apply Taylor's theorem to each of $f(x_0 + h), f(x_0 - h)$ (and the intermediate value theorem) to obtain the error estimate

$$f'(x_0) - f[x_0 - h, x_0 + h] = \frac{h^2}{6} f^{(3)}(\xi) \tag{5.36}$$

for some $\xi \in (x_0 - h, x_0 + h)$.

The approximation (5.35) can also be obtained as the result of differentiating the interpolation polynomial agreeing with f at the nodes $x_0 - h, x_0,$ and $x_0 + h$.

The contrast between the two formulas can be seen in Figure 5.10.

Example 15   Estimate $f'(0.7)$ for $f(x) = \ln x$ using (5.35)

Using the same set of steplengths as in Example 14, we get

k	$f\left[x_0 - 10^{-k}, x_0 + 10^{-k}\right]$
1	**1.4384103622589**
2	**1.4286686222028**
3	**1.42857240038993**
4	**1.42857143828945**
5	**1.42857142866326**
6	**1.42857142859665**
7	**1.42857142781949**
8	**1.42857143614616**
9	**1.42857137230834**
10	**1.42857170537525**

Again, the correct figures are shown in **bold** type.

The accuracy improves more rapidly (2 additional digits correct each time) and greater accuracy overall is achieved (10 decimal places for $k = 6$). Just as with the first-order approximation in Example 14, the accuracy is still limited and eventually the effect of the roundoff errors again takes over as we see the total error rising for small h.

Differentiation of the interpolation polynomial with the 5 nodes x_0, $x_0 \pm h$, $x_0 \pm 2h$ can be used to derive the fourth order approximation

$$f'(x_0) \approx \frac{f(x_0 - 2h) - 8f(x_0 - h) + 8f(x_0 + h) - f(x_0 + 2h)}{12h} \qquad (5.37)$$

This formula can also be obtained by Richardson extrapolation – the underlying method at the heart of Romberg integration. In this case, (5.37) would result from taking $(4f[x_0 - h, x_0 + h] - f[x_0 - 2h, x_0 + 2h])/3$.

Example 16   Use (5.37) to estimate $f'(0.7)$ for $f(x) = \ln x$

With $h = 1/10^k$, we get

k	$f'(0.7)$
1	**1.428**05826226011
2	**1.4285**713809375
3	**1.42857142856**671
4	**1.42857142857**111
5	**1.42857142856**519

We see that much greater accuracy is achieved for larger steplengths. The tendency for the error to grow when the steplength gets small is again exhibited here – but not until almost full machine accuracy has been obtained in this case.

We have made much in the last few examples of the role of the roundoff error in numerical differentiation. To examine this a little more closely, we return to the simple approximation (5.33). If h is small, then we can reasonably assume that $f(x_0)$ and $f(x_0 + h)$ have similar magnitudes and, therefore, similar roundoff errors. Suppose these are bounded by δf, and denote the computed values of f by $\hat{f}$. Then the overall error is given by

$$\left| f'(x_0) - \frac{\hat{f}(x_0 + h) - \hat{f}(x_0)}{h} \right|$$

$$\leq \left| f'(x_0) - \frac{f(x_0 + h) - f(x_0)}{h} \right| + \left| \frac{f(x_0 + h) - f(x_0)}{h} - \frac{\hat{f}(x_0 + h) - \hat{f}(x_0)}{h} \right|$$

$$\leq \left| \frac{h}{2} f''(\theta) \right| + \left| \frac{f(x_0 + h) - \hat{f}(x_0 + h)}{h} \right| + \left| \frac{f(x_0) - \hat{f}(x_0)}{h} \right|$$

$$\leq \left| \frac{h}{2} f''(\theta) \right| + \frac{2\delta}{h}$$

It is the second term of this error bound which leads to the growth of the error as $h \to 0$ illustrated in Figure 5.9. A similar effect is present for all numerical differentiation formulas.

Similar approaches to these can be used to approximate higher derivatives. For example, adding third-order Taylor expansions for $f(x_0 \pm h)$, we obtain

$$f''(x_0) \approx \frac{f(x_0 + h) - 2f(x_0) + f(x_0 - h)}{h^2} \tag{5.38}$$

which has error given by $h^2 f^{(4)}(\theta)/12$. Similarly, the five point formula is

$$f''(x_0) \approx \frac{-f(x_0 + 2h) + 16f(x_0 + h) - 30f(x_0) + 16f(x_0 - h) - f(x_0 - 2h)}{12h^2} \tag{5.39}$$

which has truncation error of order h^4.

Note that the tendency for roundoff errors to dominate the computation of higher-order derivatives is increased since the reciprocal power of h in the roundoff error estimate rises with the order of the derivative.

Example 17   Use equations (5.38) and (5.39) to estimate $f''(0.7)$ for $f(x) = \ln x$

Using (5.38):

$h = 0.1 \qquad f''(0.7) \approx 100[\ln(0.8) - 2\ln(0.7) + \ln(0.6)] = -2.061\,93$

$h = 0.01 \quad f''(0.7) \approx 10000[\ln(0.71) - 2\ln(0.7) + \ln(0.69)] = -2.041\,025$

Using (5.39):

$$h = 0.1 \qquad f''(0.7) \approx -2.03959$$
$$h = 0.01 \qquad f''(0.7) \approx -2.0408163$$

This last estimate has an error of just 5 in the last place shown.

**Exercises:
Section 5.6**

1 Use (5.33) with $h = 0.1, 0.01, \ldots, 10^{-6}$ to estimate $f'(x_0)$ for

    (*a*) $f(x) = \sqrt{x}, \quad x_0 = 1$
    (*b*) $f(x) = x^2, \quad x_0 = 2$
    (*c*) $f(x) = \ln x, \quad x_0 = 1$

2 Repeat Exercise 1 using negative steplengths $h = -10^{-k}$ for $k = 1, 2, \ldots, 6$.

3 Derive the approximation (5.35) by differentiating the interpolation polynomial which agrees with f at $x_0 - h, x_0, x_0 + h$.

4 Repeat Exercise 1 using (5.35). Compare the accuracy of your results with those of Exercises 1 and 2. Why are the results for (b) using (5.35) almost exact? Why are they *not exact*?

5 Derive the error formula (5.36) for the approximation (5.35).

6 Repeat exercise 4 using the fourth-order approximation (5.37).

7 Derive the error formula for the fourth-order approximation (5.37).

8 Estimate $f''(x_0)$ using both (5.38) and (5.39) for the same functions and steplengths as in the earlier exercises. Compare the accuracy of the two formulas.

5.7 Maxima and minima

In this section, we introduce some of the numerical techniques for finding extrema of a function f of a single variable. If f is differentiable, such points satisfy the condition $f'(x) = 0$ and so the techniques of Chapter 2 can be used to solve this equation. However, this is not always a sensible approach since there is no guarantee that we actually find an extremum (local maximum or minimum) rather than a stationary point of inflection. Usually, we seek specifically a maximum or a minimum rather than just any extremum. Again, if we just solve $f'(x) = 0$ we do not know which we have found.

We shall concentrate here on the problem of locating a *local minimum* of f. (If a local maximum is required we can apply the techniques to the minimization of $-f(x)$.) We therefore seek a point x^* which satisfies the condition

$$f(x^*) \leq f(x) \tag{5.40}$$

for all x close to x^*.

Most numerical methods for the minimization of a function of a single variable work on the principle of refining an interval in which the minimum is known to lie, and on which the function is assumed to be *unimodal* – that is, it has only one turning point within this *bracket* as the interval is called. (The problem of minimizing a function of a single variable is important in its own right, and as an important subproblem in many multivariable *optimization* techniques. Multivariate optimization is beyond the scope of this book.) The choice of a particular technique depends on what information is available – especially the availability, or otherwise, of the derivative. This applies to the first step of obtaining a suitable bracket for the minimum.

To obtain a bracket in the absence of derivative information, we require three points

$$x < y < z \text{ such that } f(y) < f(x), f(z) \tag{5.41}$$

If the derivative is available, two points suffice:

$$x < y \text{ such that } f'(x) < 0 < f'(y) \tag{5.42}$$

In general, we start with an initial guess x_0 and a steplength h which may be positive or negative. Subsequent points are then generated according to the relation

$$x_k = x_{k-1} + 2^{k-1}h \qquad (k = 1, 2, \ldots) \tag{5.43}$$

until one of the conditions (5.41) or (5.42) is satisfied by the appropriate number of successive points. Note that in (5.43) the steplength is doubled on each iteration. Even if x_0 is well removed from the true minimum, this avoids the need for a very large number of small steps – provided, of course, we are going in the right direction! This can of course result in a very wide initial bracket but, generally speaking, the methods used to refine the bracket are so much more efficient than the bracketing process itself that the overall process remains quite efficient.

We make no attempt here to describe a foolproof algorithm for bracketing a local minimum as this requires a considerable amount of (not very instructive) detail to cover all the special cases which can arise. The general principle is illustrated with examples. For a function which is known (or assumed) to be unimodal, the process is described in the following algorithms.

Algorithm 1   Bracketing without derivatives

Input   Function f, assumed to be unimodal
              Initial point x_0 and steplength h
Initial step $f_0 := f(x_0)$; $x_1 := x_0 + h$; $f_1 := f(x_1)$; $k := 1$
              If $f_1 > f_0$ then (Reverse direction)
                  $h := -h$; $x_1 := x_0 + h$; $f_1 := f(x_1)$
                  if $f_1 > f_0$ then $x_{-1} := x_0 - h$; stop.
Repeat
                  $k := k + 1$; $h := 2 * h$;
                  $x_k := x_{k-1} + h$; $f_k = f(x_k)$
until $f_k > f_{k-1}$
Output   Bracket is given by the points x_{k-2}, x_{k-1}, x_k

---

Example 18 Bracket the minimum of

$$f(x) = x + \frac{1}{x^2}$$

using Algorithm 1 with $x_0 = 2$ and $h = 0.1$

In the initial step we obtain

$$f_0 = 2.25; \quad x_1 = 2.1; \quad f_1 = 2.1 + \frac{1}{2.1^2} = 2.326\,8 > f_0.$$

This results in a change of direction and a new x_1:

$$x_1 = 1.9; \quad f_1 = 1.9 + \frac{1}{1.9^2} = 2.177\,0 < f_0.$$

In the repeat loop we then generate:

$$k = 2; \quad h = -0.2; \quad x_2 = 1.7; \quad f_2 = 2.0460 < f_1$$
$$k = 3; \quad h = -0.4; \quad x_3 = 1.3; \quad f_3 = 1.8917 < f_2$$
$$k = 4; \quad h = -0.8; \quad x_4 = 0.5; \quad f_4 = 4.5 > f_3$$

so that the minimum is bracketed by the 3 points $1.7, 1.3$ and 0.5. That is $x^* \in [0.5, 1.7]$.

Algorithm 1 relies critically on the fact that f is unimodal. It is this fact which allows us to change direction knowing that the minimum cannot lie further away in the original direction.

If the derivative is available, it is usually advantageous to use it. Since only two points are needed, it will always produce a smaller bracket (for the same initial point and steplength). In the algorithm below, we shall again assume the function is unimodal.

Algorithm 2 Bracketing with derivatives

Input   Function f, assumed to be unimodal
           Initial point x_0 and steplength $h > 0$
Initial step $df_0 := f'(x_0); k = 0$
           If $df_0 > 0$ then $h = -h$
Repeat
           $k = k + 1; x_k = x_{k-1} + h$
           $df_k = f'(x_k); h = 2 * h$
until $df_k * df_{k-1} < 0$
Output Bracket is given by x_{k-1}, x_k

Note that we have not considered the special case where $df_k * df_{k-1} = 0$ here. It is unlikely to occur in practice, and it complicates the algorithm too much if we account for it fully.

Example 19   **Bracket the minimum of**

$$f(x) = x + \frac{1}{x^2}$$

using Algorithm 2 with $x_0 = 2$ and $h = 0.1$

Here we have

$$f'(x) = 1 - \frac{2}{x^3}$$

so that $df_0 = 0.75 > 0$. We then set $h = -h = -0.1$ and get

$$k = 1; \quad x_1 = 1.9; \quad df_1 = 1 - \frac{2}{1.9^3} = 0.708\,4 > 0; \quad h = -0.2$$

$$k = 2; \quad x_2 = 1.7; \quad df_2 = 1 - \frac{2}{1.7^3} = 0.592\,9 > 0; \quad h = -0.4$$

$$k = 3; \quad x_3 = 1.3; \quad df_3 = 1 - \frac{2}{1.3^3} = 0.0897 > 0; \quad h = -0.8$$

$$k = 4; \quad x_4 = 0.5; \quad df_3 = 1 - \frac{2}{0.5^3} = -15.0 < 0$$

The bracketing interval in this case is therefore $x^* \in [0.5, 1.3]$.

We see that the final bracket in this case is just two-thirds the length of the earlier one. (Typically, it will either be one-third or two-thirds.)

We next turn to methods of obtaining the minimum to greater accuracy once a bracket is known. There are many methods available but we shall concentrate on just two. These are both based on the idea of approximating the minimum point x^* by the minimum of a local interpolation polynomial.

In the absence of derivatives, we use the *quadratic search* which, as the name suggests uses the minimum of a quadratic interpolation polynomial to estimate x^*. The interpolation polynomial agrees with the *objective function* at three points which are assumed to define a bracket for the minimum. Thus we begin with points $x_0 < x_1 < x_2$ for which

$$f(x_1) < f(x_0), f(x_2)$$

The minimum of this quadratic $\hat{x}$ will necessarily occur at a point in the interval (x_0, x_2). We use whichever 3 of the points $x_0, x_1, x_2, \hat{x}$ satisfy the conditions (5.41) for a bracket for a further iteration. First then, we require a formula for the minimum point of this interpolation quadratic.

Using the divided difference polynomial (p. 86) taking the nodes in the order x_0, x_2, x_1 we obtain

$$p(x) = f[x_0] + (x - x_0)f[x_0, x_2] + (x - x_0)(x - x_2)f[x_0, x_1, x_2] \qquad (5.44)$$

Setting $p'(x) = 0$, we obtain the minimum point

$$\hat{x} = \frac{x_0 + x_2}{2} - \frac{f[x_0, x_2]}{2f[x_0, x_1, x_2]} \qquad (5.45)$$

The situation is illustrated in Figure 5.11. The 3 interpolation points and $\hat{x}$ are marked. We see that, in this case, $\hat{x} < x_1$ and $f(\hat{x}) < f(x_1)$ so that the points $x_0, \hat{x}, x_1$ satisfy (5.41). These would be the 3 interpolation points for the next iteration.

It is also apparent from Figure 5.11 that, in the neighbourhood of the minimum, the function and its quadratic interpolant are very similar. That is to say that close to its minimum a function behaves much like a quadratic polynomial. This fact is the foundation of many optimization techniques for functions of both one and several variables.

There are special forms of the quadratic search which are designed for equally spaced nodes. Such techniques typically involve reducing the overall interval length by a factor of 1/2 on each iteration – but they typically require three new function evaluations per iteration. The method outlined above generates only one new point per iteration.

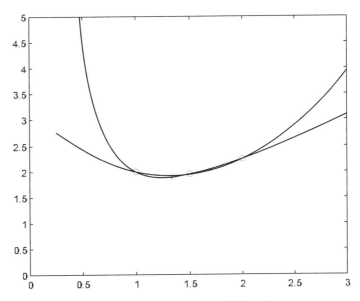

Figure 5.11 *Quadratic search iteration*

Example 20 **Perform one iteration of the quadratic search to reduce the bracket obtained in Example 18 for the minimum of $x + 1/x^2$**

The bracket was obtained from the fact that $f(1.3) < f(0.5), f(1.7)$ so that we have

$$x_0 = 0.5, \quad x_1 = 1.3, \quad x_2 = 1.7$$

with their corresponding function values $f_0 = 4.5, f_1 = 1.8917, f_2 = 2.0460$
The divided differences are then

$$
\begin{array}{llll}
0.5 & 4.5 & -2.045 & 3.03844 \\
1.7 & 2.0460 & 0.38575 & \\
1.3 & 1.8917 & &
\end{array}
$$

Using (5.45) we get

$$\hat{x} = \frac{1.7 + 0.5}{2} - \frac{-2.045}{2(3.03844)} = 1.4365$$

and $f(\hat{x}) = 1.9211$. The points x_0, $\hat{x}$, x_1 or 0.5, 1.3, 1.4365 provide a new (smaller) bracket for the next iteration.

The following MATLAB code for a basic quadratic search includes the logic used to determine the subsequent bracket.

Program   **MATLAB code for a simple quadratic search**

```
function xstar=quadmin(fcn,a,b,c,tol)
% Minimization of function fcn by quadratic search
% Initial points a<b<c such that f(b) <f(a),f(c)
% Not necessarily equally-spaced
% tol is the desired accuracy
fa=feval(fcn,a); fb=feval(fcn,b); fc=feval(fcn,c);
while c-a>tol
  fac=(fc-fa)/(c-a); fab=(fb-fa)/(b-a); fabc=(fac-fab)/(c-b);
  xs=(a+c-fac/fabc)/2; fxs=feval(fcn,xs);
  if xs<b    % Begin determination of new bracket
    if fxs<=fb
      c=b; b=xs; fc=fb; fb=fxs;
    else
      a=xs; fa=fxs;
    end
```

```
    else
      if fxs<=fb
        a=b; b=xs; fa=fb; fb=fxs;
      else
        c=xs; fc=fxs;
      end
    end
  end
  xstar=b;
```

Note that the tolerance requirement here is on the overall length of the interval. Successive iterates may be close together while the interval is still quite large.

Example 21 **Find the minimum of $x + 1/x^2$ using a quadratic search with tolerance 10^{-4}**

Using the initial bracket $[1,2]$ with its midpoint as the middle node:

```
» xmin=quadmin('mintest',1,1.5,2,1e-4)
xmin =
    1.25992105820549
```

This takes just 22 iterations which entails a total of 25 function evaluations including the initialization.

However, if we use the bracket obtained in Example 18:

```
» xmin=quadmin('mintest',0.5,1.3,1.7,1e-4)
xmin =
    1.25992105809804
```

takes 80 iterations. The reason for this is that the left-hand end-point 0.5 remains fixed for 79 iterations while the other two points get steadily closer to each other – and to the true minimum. This is due to the rapid increase in the objective function as x approaches 0.

The program listed above clearly needs a safeguard if the interval becomes 'unbalanced' in the sense that 2 of the bracketing points are very close together while the third remains distant. For example, if $(b - a) > 10(c - b)$ we could replace a with a point much closer to b and test whether that gives an improved bracket. Using such a strategy for Example 20 reduces the number of function evaluations after the initial bracket to below 20. (Of course, a similar test needs to be applied to both ends of the interval.) We do not pursue the details of this approach here. The programs presented are not intended to be robust software, but are illustrative of the basic techniques being studied.

We now turn to the *cubic search* which uses function and derivative information at 2 points to generate a cubic polynomial which has its local minimum between the 2 bracketing points.

Suppose that we have points $x_0 < x_1$ such that $f'(x_0) < 0 < f'(x_1)$ and let p be the cubic which agrees with f and f' at x_0, x_1. We can write

$$p(x) = ax^3 + bx^2 + cx + d \tag{5.46}$$

Now p has its turning points at the roots of $3ax^2 + 2bx + c$ which are

$$\frac{-b \pm \sqrt{b^2 - 3ac}}{3a} \tag{5.47}$$

and one of these – the local minimum – must lie in (x_0, x_1) since $p'(x_0) < 0 < p'(x_1)$.

If the leading coefficient $a < 0$ then the local minimum is at the smaller root which (for $a < 0$) corresponds to the + sign in (5.47). On the other hand, if $a > 0$, then the local minimum is at the larger root which again corresponds to taking the + sign. In either case therefore we set

$$\hat{x} = \frac{-b + \sqrt{b^2 - 3ac}}{3a} \tag{5.48}$$

If $f'(\hat{x}) < 0$, then we replace x_0 with $\hat{x}$; otherwise, replace x_1 with $\hat{x}$ to obtain the smaller bracket for the next iteration.

The only remaining difficulty is to obtain expressions for the coefficients. The interpolation conditions lead to the system of linear equations

$$\begin{bmatrix} x_0^3 & x_0^2 & x_0 & 1 \\ x_1^3 & x_1^2 & x_1 & 1 \\ 3x_0^2 & 2x_0 & 1 & 0 \\ 3x_1^2 & 2x_1 & 1 & 0 \end{bmatrix} \begin{bmatrix} a \\ b \\ c \\ d \end{bmatrix} = \begin{bmatrix} f(x_0) \\ f(x_1) \\ f'(x_0) \\ f'(x_1) \end{bmatrix}$$

The value of d does not affect the position of the minimum. The other coefficients are given by

$$a = X(G - 2H)$$
$$b = H - (x_1 + 2x_0)a$$
$$c = F - \left(x_1^2 + x_1 x_0 + x_0^2\right)a - (x_1 + x_0)b$$

where

$$X = 1/(x_1 - x_0)$$
$$F = X(f(x_1) - f(x_0)) = f[x_0, x_1]$$
$$G = X(f'(x_1) - f'(x_0)) = f'[x_0, x_1]$$
$$H = X(F - f'(x_0))$$

Example 22 **Perform 1 iteration of the cubic search for the minimum of** $x + 1/x^2$ **using the initial bracket obtained in Example 19**

The bracket is $[0.5, 1.3]$. The function and derivative values are

$$f_0 = 4.5, \quad f_1 = 1.8917$$
$$f_0' = -15.0, \quad f_1' = 0.0897$$

We obtain

$$X = 1/(1.3 - 0.5) = 1.25$$
$$F = 1.25(1.8917 - 4.5) = -3.2604$$
$$G = 1.25(0.0897 - (-15.0)) = 18.8621$$
$$H = 1.25(-3.2604 - (-15.0)) = 14.6745$$

and then

$$a = 1.25(18.8621 - 2(14.6745)) = -13.1086$$
$$b = 14.6745 - (1.3 + 2(0.5))(-13.1086) = 44.8243$$
$$c = -3.2604 - \left(1.3^2 + (1.3)(0.5) + 0.5^2\right)(-13.1086) - (1.3 + 0.5)44.8243$$
$$= -49.9929$$

Applying (5.48) yields

$$\dot{x} = \frac{-44.8243 + \sqrt{44.8243^2 - 3(-13.1086)(-49.9929)}}{3(-13.1086)}$$

$$= 0.9727$$

We see that $f'(\hat{x}) < 0$ and so the bracket used for the next iteration is $[0.9727, 1.3]$.

After just four iterations of the cubic search, the bracket for the example above is reduced to $[1.259921, 1.259929]$. The rapid convergence to the minimum is entirely typical of the cubic search which forms the basis of most of the so-called line search techniques used in efficient routines for the minimization of differentiable functions of several variables.

Exercises: Section 5.7

1 Find a bracket for the minimum of $f(x) = \dfrac{e^{x^3}}{x^{10}} + \dfrac{e^x}{x}$ starting with $x_0 = 1$ and steplength $h = 0.1$ without using the derivative.

2 Repeat Exercise 1, using the derivative of f.

3 Perform the first iteration of the quadratic search for the minimization of $f(x)$ in Exercise 1, using the bracket obtained there.

4 Use the quadratic search algorithm to minimize $f(x) = \dfrac{e^{x^3}}{x^{10}} + \dfrac{e^x}{x}$ with error less than 10^{-5}.

5 Perform the first iteration of the cubic search, using the bracket obtained in Exercise 2.

6 Write a program to perform the cubic search for the minimization of a function in a known bracket. Test it by finding the minimum of $x + 1/x^2$ with error less than 10^{-6}.

7 Use the cubic search to locate the minimum of the function in Exercise 1 with error less than 10^{-6}.

5.8   MATLAB functions for numerical calculus

MATLAB has functions to perform some of the tasks discussed in this chapter.

diff   can be used to assist in estimating derivatives. Its purpose is to compute differences of a MATLAB vector. If x is a MATLAB vector:

x=[x(1), x(2), . . . , x(N)]

then

» dx=diff(x);

generates the vector of differences

dx=[x(2) − x(1), x(3) − x(2),\UNICODE{0x2026}, x(N) − x(N-1)].

If we have a corresponding vector y of function values, then diff(y)./diff(x) will yield the vector of divided differences which are the simplest approximations to the first derivative.

quad, quad8   are MATLAB's basic numerical integration functions. The syntax is essentially what you would expect:

» Integral=quad('runge',0,1)
generates the result

Integral =
    0.785398125614677

using an adaptive Simpson's rule algorithm. Here runge is the function being integrated over the interval $[0, 1]$. There is an optional parameter which allows the tolerance to be set by the user. The default value is a *relative* error of 10^{-3}:

» Integral=quad8('runge',0,1)

gives the output

Integral =
    0.785398163397444

using a higher-order basic rule within an adaptive algorithm. The basic rule has error of order h^8 as opposed to Simpson's rule's h^4.

fmin is the basic function minimization function. It is based on a combination of the quadratic search described in Section 5.7 and a non-interpolation-based method, the *Golden section search*. Again the syntax is similar to that we have developed above:

» xmin=fmin('mintest',1,2)

is used to minimize the function mintest in the interval $[1,2]$. It yields the result

xmin =
    1.2599080127317

Again there is an optional parameter for the required tolerance, and for getting intermediate steps displayed.

fmins is a function for minimization of a function of several variables. It uses a geometrically-based algorithm which can be quite slow for functions of more than a very few variables. The algorithm, *Nelder and Mead's simplex* method is based on modifying an *n*-dimensional simplex (triangle, tetrahedron, . . .) according to the relative magnitudes of the function values at its vertices.

» x=fmins('rosiev',[0,0])

minimizes Rosenbrock's 'banana' function:

$$f(x,y) = 100\left(y - x^2\right)^2 + (1-x)^2$$

A contour plot of this function is shown in Figure 5.12. The contour values are $1/2$, 1, 4, 9, 16, 25, 36, 50, 100 with the smallest values on the inside. It is evident that

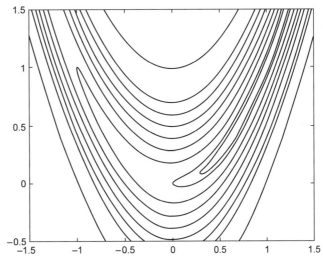

Figure 5.12   Contours of Rosenbrock's function

the function rises very sharply as we move away from a 'banana-shaped' valley centered on the curve $y = x^2$. This function is known to be a good test of multivariate minimization algorithms.

The result is the estimated minimum point

x =

1.000004 1.000011

which was achieved with a total of 146 function evaluations. The true minimum is at $(1, 1)$.

This plot was produced with the MATLAB commands

```
» x=-1.5:0.01:1.5; y=-0.5:0.01:1.5;
» [X,Y]=meshgrid(x,y);
» Z=rosie(X,Y);
» v=[1/2,1,4,9,16,25,36,50,100];
» contour(x,y,Z,v)
```

where the function rosie is defined in the m-file:

```
function z=rosie(x,y)
% Rosenbrock's 'banana' function of two variables
z=100*(y-x.^2).^2+(1-x).^2;
```

The function rosiev used earlier implements the same function with a single argument which is a vector of 2 elements.

6 Differential Equations

Aims and objectives

Many practical problems in science and engineering are appropriately modelled by *differential equations*. The objective of this chapter is to introduce the fundamental methods for their numerical solution. The methods vary with the nature of the equation to be solved. Simple initial-value problems lead to methods based on using local approximations to try to 'follow' the solution curve. These form the basis of methods used for more difficult situations. For boundary-value problems, *shooting methods* use the principle of bracketing a target, and then refining the initial conditions until the target is achieved. In the case of linear boundary-value problems, numerical differentiation formulas can be used to generate a system of linear equations for values of the solution function. This last idea provides a natural lead into Chapter 7.

6.1 Introduction and Euler's method

This chapter is concerned with the basic ideas behind the numerical solution of differential equations. We shall concentrate largely on the solution of first-order *initial-value problems*. These have the basic form

$$y' = f(x, y); \qquad y(x_0) = y_0 \qquad (6.1)$$

Later in the chapter we will consider briefly both higher-order initial-value problems (or systems of them) and the solution of *boundary-value problems* where the conditions are specified at 2 distinct points. A typical second-order 2-point boundary-value problem has the form

$$y'' = f(x, y, y'); \qquad y(a) = y_a, y(b) = y_b \qquad (6.2)$$

Most of the methods we shall consider for initial-value problems are based on a Taylor series approximation to

$$y(x_1) = y(x_0 + h_0) \qquad (6.3)$$

for some steplength h_0. The process can then be repeated for subsequent steps. An approximate value $y_1 \approx y(x_1)$ is used with a steplength h_1 to obtain an approximation $y_2 \approx y(x_2)$, and so on throughout the domain of interest. Typically, there is more than one possible derivation and explanation of the methods. Taylor series provide one. Many of the methods can also be viewed as applications of simple numerical integration rules to the integration of the function $f(x, y(x))$. The simplest method – from either of these viewpoints – is *Euler's method* for which there is also a simple graphical explanation. It is with Euler's method that we begin our study.

First, we consider Euler's method as a graphical technique. Figure 6.1 shows a slope field for the differential equation

$$y' = 3x^2y \tag{6.4}$$

which, with the initial condition $y(0) = 1$, we shall use as a basic example for much of this chapter. This particular initial-value problem

$$y' = 3x^2y; \qquad y(0) = 1 \tag{6.5}$$

is easily solved by standard techniques. However that allows us to compare our computed solution with the true one to analyze the errors.

The *slope field* (or direction field) consists of short line segments whose slopes are the slopes of solutions to (6.4) passing through those points. These slopes are computed by evaluating the right-hand side of the differential equation at a grid of points (x, y). The line segments are therefore tangent lines to solutions of the differential equation at the various points.

Using the slope field it is also easy to see the basic shape of the solution through any particular point. Try tracing a curve starting at $(0, 1)$ in Figure 6.1 following the

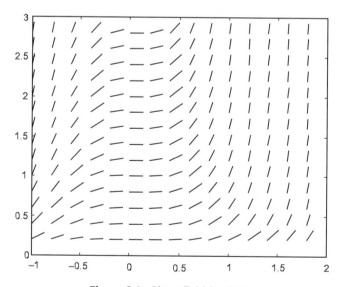

Figure 6.1 Slope field for (6.4)

slopes indicated and you will quickly obtain a good sketch of the solution to the initial-value problem (6.5).

By following a tangent line for a short distance, we obtain an approximation to the solution at a nearby point. The tangent line at that point can then be used for the next step in the solution process. This is the essence of Euler's method.

We next derive Euler's method algebraically from a Taylor approximation. For this purpose, we shall fix the steplength h. Let $x_0 = 0$ and $y_0 = 1$ as in (6.5). Also denote $x_0 + kh$ by x_k for $k = 0, 1, \ldots$ The graphical process just described is then equivalent to using a first-order Taylor approximation to $y(x_1)$

$$y(x_1) = y(x_0 + h) \approx y_0 + h y'(x_0) = y_0 + h f(x_0, y_0)$$

We denote this approximation by y_1 so that

$$y_1 = y_0 + h f(x_0, y_0) \tag{6.6}$$

is the basic step of Euler's method.

Continuing in this manner, we obtain the general Euler step:

$$y_{k+1} = y_k + h f(x_k, y_k) \tag{6.7}$$

Note 1: Apart from y_0, we use y_k to represent our approximation to the solution value $y(x_k)$. This is in contrast to our notation for interpolation and integration where we often used $f_k = f(x_k)$.

Note 2: (6.7) is not quite the generalization of (6.6) that we would like. In the first step (6.6), we have exact values to use on the right-hand side. In subsequent steps, the exact values are not available and so subsequent steps are based on approximate values. This can result in a substantial build-up of the error in our approximate solution. This build-up of error is illustrated in Figure 6.2.

The successive approximate solution points are generated by following the tangents to the different solution curves through these points. As is obvious from Figure 6.2, the approximate solution is lagging ever further behind the true solution. At each step the approximate solution follows the tangent line for $h = 0.25$. At the next step, it follows the tangent to the solution to the original differential equation that passes through this *erroneous* point. Although all the solution curves have similar behavior, those below the desired solution are all less steep at corresponding x-values and the convex nature of the curves implies that the tangent will lie below the curve resulting in further growth in the error.

We shall return to the question of the error analysis of Euler's, and other, methods shortly.

It was stated earlier that many of our methods can also be described in terms of the approximate integration of the right-hand side of the differential equation (6.1). To see this, we again consider the first step of the solution. By the fundamental theorem of calculus, we have

$$y(x_1) - y(x_0) = \int_{x_0}^{x_1} f(x, y(x)) dx \tag{6.8}$$

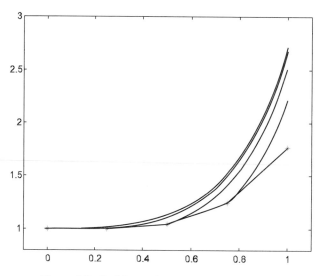

Figure 6.2   Build-up of error in Euler's method

Approximating this integral with the left-hand end-point rule, we obtain

$$y(x_1) - y(x_0) \approx (x_1 - x_0)f(x_0, y(x_0)) = hf(x_0, y(x_0))$$

which is equivalent to the approximation (6.6) derived above for Euler's method.

This suggests that alternative techniques could be based on better numerical integration rules, such as the midpoint rule or Simpson's rule. This is not completely straightforward since such formulas would require knowledge of the solution at points where it is not yet known. We return to this topic later, also.

Euler's method is easily implemented in MATLAB. The inputs required are the function f, the interval $[a, b]$ over which the solution is sought, the initial value $y_0 = y(a)$, and the number of steps to be used. The following code can be used.

Program   **MATLAB implementation of Euler's method**

```
function sol=euler1(fcn,a,b,y0,N)
%Generates Euler solution to y'=f(x,y) using N steps
%Initial condition is y(a)=y0
h=(b-a)/N;
x=a+(0:N)*h;
y(1)=y0;
for k=1:N
    y(k+1)=y(k)+h*feval(fcn,x(k),y(k));
end
sol=[x',y'];
```

The output here consists of a table of values x_k, y_k.

Note: The m-file here is called euler1.m rather than just 'euler' because there is a built-in MATLAB m-file with that name which is used for a slightly different purpose.

Example 1   Apply Euler's method to the solution of the initial-value problem (6.5) $y' = 3x^2y$; $y(0) = 1$. **Begin with** $N = 4$ **steps and repeatedly double this number of steps up to 128**

Applying the program above with $N = 4$, we use the command

```
» s4=euler1('testde',0,1,1,4)
```

to get the results

x	y
0	1.0000
0.25	1.0000
0.5	1.0469
0.75	1.2432
1.0	1.7676

We illustrate Euler's method by reproducing these results without the aid of the computer.

With $x_0 = 0$, $y_0 = 1$, $N = 4$, and $h = 1/4$, we obtain

$$y_1 = y_0 + \frac{1}{4}3x_0^2y_0 = 1$$

Then with $x_1 = 1/4, y_1 = 1$:

$$y_2 = y_1 + \frac{1}{4}3x_1^2y_1 = 1 + \frac{3}{4}\left(\frac{1}{4}\right)^2 = 1.046875$$

and, subsequently,

$$y_3 = y_2 + \frac{1}{4}3x_2^2y_2 = 1.046875 + \frac{3}{4}\left(\frac{1}{2}\right)^2 1.046875 = 1.243164$$

$$y_4 = y_3 + \frac{1}{4}3x_3^2y_3 = 1.243164 + \frac{3}{4}\left(\frac{3}{4}\right)^2 1.243164 = 1.767624$$

This is also the solution plotted in Figure 6.2. Similar tables, with more entries, are generated for the other values of N. The values for $y(1)$ obtained, and their errors, are tabulated below. (**Note**: The true solution of (6.5) is $y = \exp(x^3)$ so that $y(1) = e$.)

N	4	8	16	32	64	128
y_N	1.7676	2.1181	2.3726	2.5312	2.6207	2.6684
$\lvert Error \rvert$	0.9507	0.6002	0.3456	0.1870	0.0975	0.0498

We see that the errors are steadily, if slowly being reduced. The first few solutions are plotted, along with the true solution, in Figure 6.3.

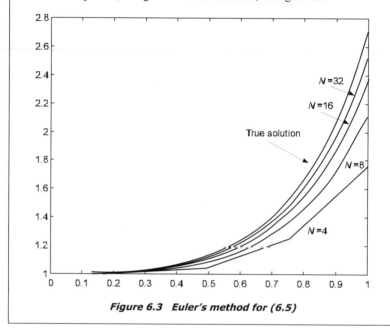

Figure 6.3   Euler's method for (6.5)

The ratios of the successive final errors in Example 1 are approximately 0.63, 0.58, 0.54, 0.52 and 0.51 which seem to be settling down close to 1/2 which is, of course, the same factor by which the steplength h is reduced. This suggests that the overall error in Euler's method is of order h; that is, the error is $O(h)$.

We turn now to the analysis of this truncation error. As we have already observed, there are 2 components to this error. There is a contribution due to the straight line approximation used at each step. There is also a significant contribution resulting from the earlier errors. Their effect is that the linear approximation is not tangent to the required solution but to another solution of the differential equation passing through the current point. These 2 contributions are called the *local* and *global* truncation errors.

The situation is illustrated in Figure 6.4 for step 2 of Euler's method.

In this case, the local contribution to the error at step 2 is relatively small, with a much larger contribution coming from the effect of the (local and global) error in step 1.

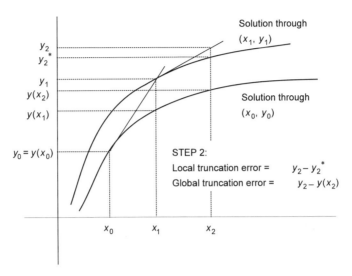

Figure 6.4 *Global and local truncation error*

For step 1, we have

$$y_1 = y_0 + hf(x_0, y_0)$$

while Taylor's theorem gives us

$$y(x_1) = y_0 + hy_0' + \frac{h^2}{2}y''(\xi) = y_0 + hf(x_0, y_0) + \frac{h^2}{2}y''(\xi)$$

It follows that the local truncation error in the approximation is

$$|y_1 - y(x_1)| = \frac{h^2}{2}|y''(\xi)| \le Mh^2$$

where M is a bound on the second derivative of the solution. A similar local truncation error will occur at each step of the solution process.

We see that the local truncation error for Euler's method is $O(h^2)$ yet we observed that the global truncation error appears to be just $O(h)$.

By the time we have computed our estimate y_N of $y(b)$, we have committed N of these truncation errors from which we may conclude that the global truncation error has the form

$$E = NMh^2 = M(b - a)h$$

since $Nh = b - a$. Thus the global truncation error is indeed of order h.

The 'derivation' above is not rigorous, but it does lead to the correct result in a reasonably intuitive manner. The stated error is the leading contribution to the rigorously defined global truncation error which is of order h. (The rigorous derivation is more complicated than is appropriate here.)

In this error analysis, we have completely ignored the effect of roundoff errors. These too will tend to accumulate and their effect can be analyzed in a manner similar to that for the global truncation error. The result is that the global roundoff error *bound* is *inversely* proportional to the steplength h. As with numerical differentiation in Chapter 5, this places a restriction on the accuracy that can be achieved. However, for systems such as MATLAB, working with IEEE double-precision arithmetic, the effect of roundoff error is typically *much* smaller than the global truncation error, unless we are seeking very high accuracy and therefore using very small steplengths.

Clearly, Euler's method is not itself going to be sufficient to generate accurate solutions to differential equations. It does, however, provide the basis for much of what follows in the next two sections.

Exercises: Section 6.1

Exercises 1–6 and 8 concern the differential equation

$$y' = x/y$$

1   Find the general solution of the differential equation.
2   For the initial condition $y(0) = 3$, use Euler's method with steps $h = 1, 1/2$ and $1/4$ to approximate $y(1)$.
3   Use Euler's method to solve the initial-value problem of Exercise 2 over $[0, 4]$ with $N = 10, 20, 50, 100, 200$ steps.
4   Tabulate the errors in the approximate values of $y(4)$ in Exercise 3. Verify that their errors appear to be $O(h)$.
5   Using negative steplengths solve the initial-value problem in Exercise 2 over $[-4, 0]$ to obtain a graph of its solution over $[-4, 4]$.
6   Repeat Exercise 5 for different initial conditions: $y(0) = 1, 2, 3, 4, 5$ and plot the solutions on the same axes.
7   Use Euler's method with steplengths $h = 10^{-k}$ for $k = 1, 2, 3$ to solve the initial-value problem $y' = x + y^2$ with $y(0) = 0$ on $[0, 1]$. Tabulate the results for $x = 0, 0.1, 0.2, \ldots, 1$ and graph the solutions.
8   (More challenging) Use Euler's method to solve the initial-value problems with initial conditions $y(k) = 0$ over the interval $[k, 10]$ for $k = 1, 2, 3, 4, 5$. Plot these solutions on the same axes. (**Note** The slope of the solution is infinite at $x = k$. The differential equation can be rewritten in the form $dx/dy = y/x$ for these cases. This can be solved in the same way as before – except that we do not know the y-interval over which we must compute the solution.)

6.2   Runge–Kutta methods

The basic idea of the *Runge–Kutta methods* is to use additional points between x_k and x_{k+1} which allow the resulting approximation to agree with more terms of the

Taylor series. Another interpretation of this is that we are able to use higher-order numerical integration methods to approximate

$$y(x_{k+1}) - y(x_k) = \int_{x_k}^{x_{k+1}} f(x,y)dx$$

The general derivation of Runge–Kutta methods is somewhat complicated. We shall illustrate the process with examples of second-order Runge–Kutta methods which are derived from a second-order Taylor series expansion. Higher-order, especially fourth-order, Runge–Kutta methods are commonly used in practice. The basic ideas behind these are similar to those presented. We shall discuss the most commonly used fourth-order Runge–Kutta method in some detail without deriving it in detail.

Consider the second-order Taylor expansion about the point (x_0, y_0) using a steplength h:

$$y(x_1) \approx y_0 + hy_0' + \frac{h^2}{2}y_0''$$

from which we obtain the approximate value

$$y_1 = y_0 + hf(x_0, y_0) + \frac{h^2}{2}y_0'' \tag{6.9}$$

This formula has error of order $O(h^3)$. However, we need an approximation for y_0''. Since this term is multiplied by h^2, it follows that an approximation to y_0'' which is itself accurate to order $O(h)$ will preserve the overall error in (6.9) at $O(h^3)$.

It will be helpful to introduce a little notation here. Denote the slope $y_0' = f(x_0, y_0)$ by k_1. Also let $\alpha \in [0,1]$ and consider an 'Euler step' of length αh. We get

$$y(x_0 + \alpha h) \approx y_0 + \alpha h k_1$$

and denote by k_2 the slope at the point $(x_0 + \alpha h, y_0 + \alpha h k_1)$ so that $k_2 = f(x_0 + \alpha h, y_0 + \alpha h k_1)$. The simplest divided difference estimate of y_0'' is then given by

$$\begin{aligned}
y_0'' &\approx \frac{y'(x_0 + \alpha h) - y'(x_0)}{\alpha h} \\
&\approx \frac{hf(x_0 + \alpha h, y_0 + \alpha h k_1) - f(x_0, y_0)}{\alpha h} \\
&= \frac{k_2 - k_1}{\alpha h}
\end{aligned} \tag{6.10}$$

Substituting this approximation into (6.9), we obtain

$$\begin{aligned}
y_1 &= y_0 + hk_1 + \frac{h^2}{2}\frac{k_2 - k_1}{\alpha h} \\
&= y_0 + h\left[k_1\left(1 - \frac{1}{2\alpha}\right) + \frac{k_2}{2\alpha}\right]
\end{aligned} \tag{6.11}$$

which is the general form of the Runge–Kutta second-order formulas. The corresponding formula for the n-th step in the solution process is

$$y_{n+1} = y_n + h\left[k_1\left(1 - \frac{1}{2\alpha}\right) + \frac{k_2}{2\alpha}\right] \qquad (6.12)$$

where now

$$k_1 = f(x_n, y_n)$$
$$k_2 = f(x_n + \alpha h, y_n + \alpha h k_1) \qquad (6.13)$$

Because the approximation (6.10) has error of order $O(h)$, it follows that the local truncation error in (6.11) is $O(h^3)$ as desired. We shall return to the error analysis of these methods shortly.

There are three important special cases of (6.12) which correspond to choosing $\alpha = 1/2$, 1 or 2/3.

$\alpha = 1/2$: **corrected Euler, or midpoint, method**

$$y_{n+1} = y_n + hk_2 \qquad (6.14)$$

where $k_2 = f(x_n + h/2, y_n + hk_1/2)$. This method is essentially the application of the midpoint rule for integration of $f(x,y)$ over the current step with the (approximate) value at the midpoint being obtained by a preliminary (half-) Euler step.

$\alpha = 1$: **modified Euler method**

$$y_{n+1} = y_n + \frac{h}{2}(k_1 + k_2) \qquad (6.15)$$

with $k_2 = f(x_n + h, y_n + hk_1)$. This method corresponds to using the trapezoid rule to estimate the integral where a preliminary (full) Euler step is taken to obtain the (approximate) value at x_{n+1}.

$\alpha = 2/3$: **Heun's method**

$$y_{n+1} = y_n + \frac{h}{4}(k_1 + 3k_2) \qquad (6.16)$$

with $k_2 = f(x_n + 2h/3, y_n + 2hk_1/3)$. Heun's method does not correspond to any of the standard numerical integration formulas.

The corrected Euler, or midpoint, and modified Euler methods are illustrated in Figure 6.5.

In each case of Figure 6.5, the second curve represents the solution to the differential equation passing through the appropriate point. For the corrected Euler this 'midpoint' slope k_2 is then applied for the full step from x_0 to x_1. For the modified Euler method, the slope used is the average of the slope k_1 at x_0 and the (estimated) slope at x_1, k_2. It is clear that both these methods result in significantly improved estimates of the average slope of the true solution over $[x_0, x_1]$ than Euler's method, which just uses k_1.

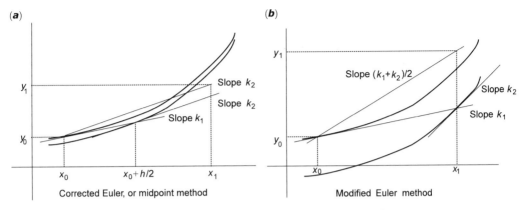

Figure 6.5 *Second-order Runge–Kutta methods*

We have already observed that the local truncation error for all these 'second-order' Runge–Kutta methods is $O(h^3)$. So, why are they called 'second-order' methods? In just the same way as for Euler's method, by the time we have computed our approximate value of $y(b) = y(a + Nh)$, we have committed N of these local truncation errors. The effect is that the global truncation error is of order $O(Nh^3)$ which is $O(h^2)$ since $Nh = b - a$ is constant. (As for Euler's method, this is not a rigorous argument, but it explains the situation adequately for our present purpose.)

It follows that reducing the steplength by a factor of 1/2 should now result in approximately a 75% improvement in the final answer. That is the error should be reduced by a factor of (about) 1/4. We shall verify this with examples.

Example 2 Use the second-order Runge–Kutta methods (6.14), (6.15) and (6.16) with $N = 2$ to solve the initial-value problem $y' = 3x^2 y$; $y(0) = 1$ on $[0, 1]$

For the Corrected Euler (midpoint) (6.14) method we obtain the results below. The true solution is tabulated for comparison:

k	Corrected Euler	True solution value
0	$k_1 = f(0, 1) = 0$ $k_2 = f(1/4, 1) = 3/16$ $y_1 = 1 + (1/2)(3/16) = 1.09375$	$\exp(1/8) = 1.133148$
1	$k_1 = f(1/2, 1.09375) = 0.8203125$ $k_2 = f(3/4, 1.298828) = 2.191772$ $y_1 = 1.09375 + (1/2)(2.191772) = 2.189636$	$\exp(1) = 2.718282$

The other two methods yield the results tabulated here in somewhat less detail:

k	Modified Euler	Heun's method
0	$k_1 = f(0,1) = 0$ $k_2 = f(1/2,1) = 3/4$ $y_1 = 1 + (1/4)(0 + 3/4) = 1.1875$	$k_1 = 0$ $k_2 = 1/3$ $y_1 = 1.125$
1	$k_1 = f(1/2, 1.1875) = 0.890625$ $k_2 = f(1, 1.632813) = 4.898439$ $y_1 = 2.634766$	$k_1 = 0.84375$ $k_2 = 2.929688$ $y_1 = 2.329102$

All of these methods are easily programmed. (See the Exercises.) They can be applied to this same example with more steps.

Example 3 Apply the three Runge–Kutta methods (6.14), (6.15) and (6.16) to the same initial-value problem with $N = 10$ and $N = 100$. Compare their errors and verify that the global truncation error is $O(h^2)$

For $N = 10$, the following MATLAB commands produce the results tabulated below.

```
» sc=correuler('testde',0,1,1,10);
» sm=modeuler('testde',0,1,1,10);
» sh=heun('testde',0,1,1,10);
```

where the m-files implementing the various methods have fairly obvious names. (These are not built-in MATLAB functions.)

x	Corrected	Modified	Heun
0	1.0000	1.0000	1.0000
0.1	1.0008	1.0015	1.0010
0.2	1.0075	1.0090	1.0080
0.3	1.0265	1.0289	1.0273
0.4	1.0648	1.0681	1.0659
0.5	1.1310	1.1357	1.1326
0.6	1.2375	1.2442	1.2397
0.7	1.4028	1.4128	1.4061
0.8	1.6569	1.6722	1.6619
0.9	2.0505	2.0749	2.0585
1.0	2.6732	2.7138	2.6865

For this particular example, the modified Euler method appears to be performing somewhat better than the other two. The errors at $x = 1$ are

Corrected	Modified	Heun
-0.0451	-0.0045	-0.0318

For comparison, the error in Euler's method for $N = 10$ is -0.5069, so that we see a substantial improvement resulting from using the second-order methods.

For $N = 100$, we of course do not tabulate all the results. The errors at $x = 1$ are now

Corrected	Modified	Heun
-0.00054	-0.00004	-0.00037

while that for Euler's method is -0.0634.

With the increase in N by a factor of 10 we should expect the error in the first-order Euler's method to be reduced by approximately this same factor. The errors in the second-order methods should be reduced by a factor of about 10^2. It is apparent that *approximately* this level of improvement has indeed been achieved.

For the particular equation in Example 3, it appears that the modified Euler method (which uses an approximate trapezoid rule to integrate the slope) is performing best. This should not be taken as an indicator of the relative merits of these methods in general. In this particular case, it is probably due to the fact that the function is sharply convex (concave up), so that using slope estimates from further to the right is likely to be beneficial.

One of the most widely used Runge–Kutta methods is the classical fourth-order formula, RK4

$$y_{n+1} = y_n + \frac{h}{6}[k_1 + 2(k_2 + k_3) + k_4] \tag{6.17}$$

where the various slope estimates are

$$k_1 = f(x_n, y_n)$$
$$k_2 = f(x_n + h/2, y_n + hk_1/2)$$
$$k_3 = f(x_n + h/2, y_n + hk_2/2)$$
$$k_4 = f(x_n + h, y_n + hk_3)$$

The derivation of this result, and of the fact that its global truncation error is $O(h^4)$, are omitted here. The basic idea is again that of obtaining approximations which agree with more terms of the Taylor expansion.

In keeping with the earlier methods we can also view (6.17) as an approximate application of Simpson's rule to the integration of the slope over $[x_n, x_{n+1}]$. Here k_1 is the slope at x_n, k_2 and k_3 are both estimated slopes at the midpoint $x_n + h/2$, and k_4 is then an estimate of this slope at $x_n + h = x_{n+1}$. With this interpretation, (6.17) is equivalent to Simpson's rule where the value at the midpoint is replaced by the average of the 2 estimates. This interpretation of the classical RK4 formula also lends some credence to the claim that it *is* a fourth-order method. We already know that Simpson's rule has an error of order $O(h^4)$, and we have already seen that the second-order Runge–Kutta methods (6.14) and (6.15) have errors of the same order as their corresponding integration rules.

Program

MATLAB code for the classical Runge–Kutta fourth-order method RK4 (6.17)

```
function sol=RK4(fcn,a,b,y0,N)
%Generates RK4 solution to y'=f(x,y) using N steps
%Initial condition is y(a)=y0
h=(b-a)/N;
x=a+(0:N)*h;
y(1)=y0;
for k=1:N
  k1=feval(fcn,x(k),y(k));
  k2=feval(fcn,x(k)+h/2,y(k)+h*k1/2);
  k3=feval(fcn,x(k)+h/2,y(k)+h*k2/2);
  k4=feval(fcn,x(k)+h,y(k)+h*k3);
  y(k+1)=y(k)+h*(k1+2*(k2+k3)+k4)/6;
end
sol=[x',y'];
```

Example 4   Apply RK4 to our usual example with both $N = 10$ and $N = 100$

Using $N = 10$ the command

    » rk=rk4('testde',0,1,1,10)

produces the results tabulated.

x	RK4 solution	True solution
0	1.0000	1.0000
0.1	1.0010	1.0010
0.2	1.0080	1.0080
0.3	1.0274	1.0274
0.4	1.0661	1.0661
0.5	1.1331	1.1331

x	*RK4 solution*	*True solution*
0.6	1.2411	1.2411
0.7	1.4092	1.4092
0.8	1.6686	1.6686
0.9	2.0730	2.0730
1.0	2.7182	2.7183

It is apparent that this method is producing results which are very close to the correct values. The error in $y(1)$ is just -4.8×10^{-5}.

With $N = 100$, this error is reduced to only -5.0×10^{-9}, which reflects the reduction by a factor of 10^4 which we should anticipate for a fourth-order method when the number of steps is increased by a factor of 10.

We have seen in this section that higher-order Runge–Kutta methods can give high-accuracy answers with small numbers of steps – at least when the differential equation is not too difficult. There is an enormous theory of numerical methods for situations where these methods are inadequate. This involves what are called *stiff* differential equations and the concepts of *stability* and what is called *A-stability*.

These topics are well beyond the scope of this text. Some of the techniques used have much in common with the adaptive methods for numerical integration. One widely used example is the Runge–Kutta–Fehlberg method which uses RK4 in conjunction with a fifth-order method to make adaptive alterations to the steplength as the solution proceeds.

Exercises: Section 6.2

Exercises 1–5 and 8 concern the differential equation

$$y' = x/y$$

1   For the initial condition $y(0) = 3$, use the corrected Euler's method with steps $h = 1, 1/2$ and $1/4$ to approximate $y(1)$.

2   Repeat Exercise 1 for the modified Euler and Heun methods.

3   Write MATLAB programs to implement the three second-order Runge–Kutta methods we have discussed. Use them to solve the initial-value problem of Exercises 1 and 2 over $[0, 4]$ with $N = 10, 20, 50, 100$ steps.

4   Tabulate the errors in the approximate values of $y(4)$ in Exercise 3. Verify that their errors appear to be $O(h^2)$. Which of the methods appears best for this problem?

5   Repeat Exercise 3 for different initial conditions: $y(0) = 1, 2, 3, 4, 5$ and plot the solutions for $N = 100$ on the same axes.

6   Show that if $f(x, y)$ is a function of x alone, then the modified Euler method is just the trapezoid rule with step h.

7   Use the classical Runge–Kutta RK4 method with steplengths $h = 10^{-k}$ for $k = 1, 2, 3$ to solve the initial-value problem $y' = x + y^2$ with $y(0) = 0$ on $[0, 1]$. Tabulate the results for $x = 0, 0.1, 0.2, \ldots, 1$ and graph the solutions.

8 Repeat Exercise 7 for the initial-value problem of Exercise 1. Verify that the errors are of order $O(h^4)$.

9 Show that if $f(x, y)$ is a function of x alone, then RK4 is just Simpson's rule with step $h/2$.

10 Solve the differential equation $y' = -x \tan y$ with $y(0) = \pi/6$ for $x \in [0, 1]$. Compare your solution at the points $0 : 0.1 : 1$ with the values obtained using RK4 with $h = 0.1$ and $h = 0.05$.

6.3 Multistep methods

In this section, we consider methods which, like the Runge–Kutta methods, use more than one point to compute the estimated value of y_{k+1}. This time, however, all the information is taken from tabulated points.

The simplest example is based on a second-order Taylor expansion

$$y_{n+1} = y_n + hy_n' + \frac{h^2}{2}y_n'' = y_n + hf(x_n, y_n) + \frac{h^2}{2}y_n'' \tag{6.18}$$

Using a simple backward difference approximation to the second derivative, we have

$$y_n'' \approx \frac{y'(x_n) - y'(x_n - h)}{h} \approx \frac{f(x_n, y_n) - f(x_{n-1}, y_{n-1})}{h}$$

and substituting this into (6.18) we obtain the *2-step formula*

$$y_{n+1} = y_n + hf(x_n, y_n) + \frac{h^2[f(x_n, y_n) - f(x_{n-1}, y_{n-1})]}{2h}$$

$$= y_n + \frac{h}{2}(3f_n - f_{n-1}) \tag{6.19}$$

where we have abbreviated $f(x_n, y_n)$ to just f_n. This particular formula is the 2-step *Adams–Bashforth* method. The formula (6.19) agrees with the Taylor expansion as far as the second-order term and so has local truncation error of order $O(h^3)$, and consequently, the global truncation error is $O(h^2)$.

We shall consider only the class of multistep methods known as Adams methods. These include the Adams–Bashforth methods as one of their two important subclasses. We shall introduce the other class shortly.

An N-step Adams method uses values from N previous points in order to estimate y_{n+1}. The general formula is therefore

$$y_{n+1} = y_n + h\sum_{k=0}^{N}\beta_k f_{n+1-k} \tag{6.20}$$

The coefficients β_k are chosen to give the maximum order of agreement with the Taylor expansion. We shall see later that, like the Runge–Kutta methods, these Adams methods can be viewed as the application of numerical integration rules – this time incorporating nodes from outside the immediate interval.

Note that in (6.20) the coefficient β_0 multiplies the slope $f_{n+1} = f(x_{n+1}, y_{n+1})$. Thus if $\beta_0 \neq 0$, the right-hand side of (6.20) depends on the very quantity y_{n+1} which we are trying to estimate. Such a formula may not seem initially very promising. Adams methods with $\beta_0 \neq 0$ are called *Adams–Moulton* methods. They give an *implicit* formula for y_{n+1} which could be solved iteratively. However, we shall see an alternative way of using the Adams–Moulton methods shortly.

The cases where $\beta_k = 0$ are the general N-step Adams–Bashforth methods which provide *explicit* formulas for y_{n+1}.

Next, we derive the Adams–Moulton method which agrees with the Taylor expansion up to third order. We have

$$y(x_{n+1}) \approx y_n + hy_n' + \frac{h^2}{2}y_n'' + \frac{h^3}{6}y_n''' \tag{6.21}$$

and we can again use difference approximations to the higher derivatives. Since the Adams–Moulton formulas are implicit, we can use the (better) symmetric approximations to the derivatives:

$$y_n'' \approx \frac{y_{n+1}' - y_{n-1}'}{2h}$$
$$= \frac{f_{n+1} - f_{n-1}}{2h}$$

and

$$y_n''' \approx \frac{y_{n+1}' - 2y_n' + y_{n-1}'}{h^2}$$
$$= \frac{f_{n+1} - 2f_n + f_{n-1}}{h^2}$$

(see Section 5.6). Substituting these into (6.21), we obtain

$$y_{n+1} = y_n + hf_n + \frac{h^2}{2}\frac{f_{n+1} - f_{n-1}}{2h} + \frac{h^3}{6}\frac{f_{n+1} - 2f_n + f_{n-1}}{h^2}$$
$$= y_n + \frac{h}{12}(5f_{n+1} + 8f_n - f_{n-1}) \tag{6.22}$$

which is the 2-step Adams–Moulton formula.

The 2-step Adams–Moulton formula (6.23) agrees with the third-order Taylor expansion and so has local truncation error $O(h^4)$ and a corresponding global truncation error of order $O(h^3)$. This illustrates the potential benefit of the Adams–Moulton methods. The N-step Adams–Moulton method has error of order one greater than the corresponding N-step Adams–Bashforth method.

The second-order Adams–Moulton formula is the 1-step method which corresponds to the implicit trapezoid rule

$$y_{n+1} = y_n + \frac{h}{2}(f_{n+1} + f_n) \tag{6.23}$$

So far, we have completely ignored one apparent major problem with these multistep methods. How do we start them? To use the 2-step Adams–Bashforth formula (6.19) to obtain y_1 appears to require knowledge of y_{-1} which would not be available. For higher-order methods using more steps, this difficulty is increased. The usual solution to this problem is that a small number of steps of a Runge–Kutta method of the desired order are used to generate enough values to allow the Adams methods to proceed.

For example, we could use the modified Euler method (a second-order Runge–Kutta method) to generate y_1 after which the second-order (2-step) Adams–Bashforth method can be used to generate $y_2, y_3, \dots$. For a fourth order Adams method, we would use RK4 to generate y_1, y_2, y_3 after which we have enough points to continue with the Adams method.

The advantage is that, once the Adams methods are started, subsequent points are generated without the need for any of the intermediate points required by the Runge–Kutta methods. Therefore, for example, the second-order Adams–Bashforth method would use only about half the computational effort of a second-order Runge–Kutta method with the same basic steplength.

Example 5   **Use the 2-step Adams–Bashforth method to solve the initial-value problem $y' = 3x^2 y$; $y(0) = 1$ on $[0, 1]$ with $h = 1/4$**

We begin by using a second-order Runge–Kutta method to get y_1. The modified Euler method yields

$$y_1 = 1.0234375$$

The Adams–Bashforth method can now be used to generate the remaining values as follows:

x	y	f
0	1	0
1/4	1.023438	0.191895
1/2	$1.023438 + (1/8)(3(0.191895) - 0) = 1.095399$	0.821549
3/4	$1.095399 + (1/8)(3(0.821549) - 0.191895) = 1.379493$	2.327894
1	$1.379493 + (1/8)(3(2.327894) - 0.821549) = 2.149758$	6.449274

Example 6 **Use the Adams–Bashforth 2-step method to solve the same initial-value problem as in Example 5 using $N = 10$ steps**

This is easy to program. The m-file used was as follows:

```
function s=ab2(fcn,a,b,y0,N)
% Adams-Bashforth 2-step (2nd order) solution of
% y'=fcn(x,y) on [a,b] with y(a)=y0 and N steps
h=(b-a)/N;
x=a+(0:N)*h;
y(1)=y0; f(1)=feval(fcn,x(1),y(1));
% First step uses Modified Euler to generate y(2)
k1=f(1);
k2=feval(fcn,x(1)+h,y(1)+h*k1);
y(2)=y(1)+h*(k1+k2)/2;
f(2)=feval(fcn,x(2),y(2));
% Now use 2-step formula for rest of the interval
for k=2:N
        y(k+1)=y(k)+h*(3*f(k)-f(k-1))/2;
        f(k+1)=feval(fcn,x(k+1),y(k+1));
end
s=[x',y'];
```

The choice of the modified Euler method to generate y_1 (or y(2) in the MATLAB code) is somewhat arbitrary, any of the second-order Runge–Kutta methods could have been used.

The results are included in the second column of Table 6.1.

For comparison, the third column of Table 6.1 shows the results obtained using the modified Euler method (Example 3) with the same number of steps. These 2 methods have global truncation errors of the same order $O(h^2)$. The explicit Adams–Bashforth method appears much inferior. *But* we must note that each step of AB2 requires only one new evaluation of the right-hand side, whereas the modified Euler method (or any second-order Runge–Kutta method) requires 2 such evaluations per step.

It would therefore be fairer to compare the 2-step Adams–Bashforth method using $N = 20$ with the modified Euler method using $N = 10$. These results form the third set of output. As expected the error is reduced by (approximately) a factor of $1/4$ compared to $N = 10$. Although the error is still greater than that for the modified Euler method, it is now comparable with those for the other second-order Runge–Kutta methods in Example 3.

What is the role of the Adams–Moulton methods? They appear to have a further difficulty due to being implicit methods. Their primary use is in conjunction with

Table 6.1   Results for Adams methods for the initial-value problem of Example 6

x	*AB2, N* = 10	*ModEuler, N* = 10	*AB2, N* = 20	*ABM23, N* = 10
0	1.0000	1.0000	1.0000	1.0000
0.1	1.0015	1.0015	1.0008	1.0015
0.2	1.0060	1.0090	1.0071	1.0085
0.3	1.0226	1.0289	1.0258	1.0279
0.4	1.0580	1.0681	1.0636	1.0666
0.5	1.1204	1.1357	1.1293	1.1338
0.6	1.2210	1.2442	1.2351	1.2418
0.7	1.3768	1.4128	1.3995	1.4099
0.8	1.6145	1.6722	1.6523	1.6692
0.9	1.9782	2.0749	2.0440	2.0728
1.0	2.5443	2.7138	2.6640	2.7153

Adams–Bashforth methods as *predictor–corrector* pairs. The basic idea is that an explicit method is used to estimate y_{n+1}, after which this estimate can be used in the right-hand side of the Adams–Moulton (implicit) formula to obtain an improved estimate. Thus the Adams–Bashforth method is used to *predict* y_{n+1} and the Adams–Moulton method is then use to *correct* this estimate.

Example 7   Use an Adams predictor–corrector method to solve the usual example problem $y' = 3x^2y$; $y(0) = 1$ on $[0, 1]$. Specifically, use the two 2-step methods (6.19) and (6.22) with $N = 10$ steps

The code for the 2-step Adams–Bashforth method is readily modified to add the corrector step. The basic loop becomes

```
for k=2:N
    y(k+1)=y(k)+h*(3*f(k)-f(k-1))/2; % predictor
    f(k+1)=feval(fcn,x(k+1),y(k+1));
    y(k+1)=y(k)+h*(5*f(k+1)+8*f(k)-f(k-1))/12; % corrector
    f(k+1)=feval(fcn,x(k+1),y(k+1));
end
```

The results are shown as the final column of Table 6.1.

This time we see that the predictor–corrector method has provided a smaller error than the modified Euler method using the same number of points. This comparison is a fair one since both methods entail two new evaluations of the slope per step. The Adams predictor–corrector approach seems to justify our investigation of multistep methods.

We now return to the issue of determining the coefficients for Adams methods. This will re-emphasize the connection between differential equations and numerical integration formulas.

If $f(x,y)$ has no explicit dependence on y then it follows that

$$y_{n+1} - y_n = \int_{x_n}^{x_{n+1}} f(x)dx = h \int_0^1 f(x_n + ht)dt$$

We can therefore seek values of the coefficients in (6.20) so that the corresponding quadrature formula has maximal degree of precision. (**Note:** The integration formula includes some nodes from outside the range of integration corresponding to $t = -1, -2, \ldots$.)

For a 3-step Adams–Bashforth formula we therefore seek β_1, β_2, β_3 so that the quadrature formula

$$\int_0^1 F(t)dt \approx \beta_1 F(0) + \beta_2 F(-1) + \beta_3 F(-2)$$

is exact for all quadratic functions F. Imposing this condition for $F(t) = 1, t, t^2$ in turn we get the equations

$$F(t) = 1 \;:\; \beta_1 + \beta_2 + \beta_3 = 1$$
$$F(t) = t \;:\; -\beta_1 - 2\beta_2 = 1/2$$
$$F(t) = t^2 \;:\; \beta_2 + 4\beta_3 = 1/3$$

Adding the last two of these yields $\beta_3 = 5/12$, and thence we obtain $\beta_2 = -4/3$, $\beta_1 = 23/12$. Therefore the 3-step Adams–Bashforth formula is

$$y_{n+1} = y_n + \frac{h(23f_n - 16f_{n-1} + 5f_{n-2})}{12} \tag{6.24}$$

(**Note:** The sum of the coefficients in such a formula must be unity, which provides a useful and simple check.)

The 3-step formula (6.24) has fourth-order local truncation error so that its global truncation error is of order $O(h^3)$.

The corresponding 3-step (fourth-order) Adams–Moulton formula is based on the quadrature formula

$$\int_0^1 F(t)dt \approx \beta_0 F(1) + \beta_1 F(0) + \beta_2 F(-1) + \beta_3 F(-2)$$

Using a similar approach to that above we obtain

$$y_{n+1} = y_n + \frac{h(9f_{n+1} + 19f_n - 5f_{n-1} + f_{n-2})}{24} \tag{6.25}$$

A simple modification of the code used in Examples 6 and 7 could be used to implement a 3-step predictor corrector method. The corrector formula has $O(h^4)$ truncation error and so we might compare its performance with the fourth-order Runge–Kutta formula.

Example 8   **Compare the performance of the 3-step Adams predictor–corrector method with the fourth-order Runge–Kutta method on our usual example problem**

The results for RK4 with $N = 10$ steps were computed in Example 4 (p. 184). The table of output below shows these and the results for the predictor–corrector method using $N = 10$ and $N = 20$ steps.

x	RK4, $N = 10$	ABM34, $N = 10$	ABM34, $N = 20$	True solution
0	1.000000	1.000000	1.000000	1.000000
0.1	1.001000	1.001000	1.001000	1.001001
0.2	1.008032	1.008032	1.008033	1.008032
0.3	1.027368	1.027383	1.027370	1.027368
0.4	1.066092	1.066131	1.066096	1.066092
0.5	1.133148	1.133214	1.133154	1.133148
0.6	1.241102	1.241190	1.241110	1.241102
0.7	1.409168	1.409250	1.409175	1.409169
0.8	1.668621	1.668601	1.668623	1.668625
0.9	2.072993	2.072581	2.072969	2.073007
1.0	2.718233	2.716599	2.718129	2.718282

Now, each step of the RK4 algorithm requires 4 new evaluations of the right-hand side, whereas each step of the predictor–corrector method uses just 2. The 3-step Adams method with $N = 20$ is therefore comparable with RK4 using $N = 10$. Their performance is also comparable. For this particular example the RK4 error is slightly smaller.

With the numbers of steps increased to $N = 100$ for RK4 and 200 for ABM34, the final errors in $y(1)$ are 5.0×10^{-9} and 2.2×10^{-8}, respectively. Again, the two methods produce similar errors for similar computational effort.

The performance of a predictor–corrector pair can sometimes be improved by repeating the corrector step. For the particular example used here, this policy shows no significant benefit.

Exercises: Section 6.3

Exercises 1–6 and 10–12 concern the initial-value problem $y' = x/y$, $y(0) = 3$

1   Apply the 2-step Adams–Bashforth method AB2 with $N = 4$ on $[0, 1]$. Use the modified Euler method for the first step.

2   Compare AB2 using $N = 20$ steps with the modified and corrected Euler methods using $N = 10$ steps on $[0, 4]$. (The computational effort is comparable in all these.)

3   Repeat Exercise 2 for $N = 40, 100, 200$ for AB2 and $N = 20, 50, 100$ for the Runge–Kutta methods. Verify that the truncation error of AB2 is second-order.

4　Use the predictor–corrector method with AB2 as predictor and the second-order Adams–Moulton formula AM2 (implicit trapezoid rule) as corrector. Take $N = 10$ and compare the results with those of Exercise 2.

5　Repeat Exercise 4 using the third order Adams–Moulton formula as corrector.

6　Use $N = 20$, 50, 100 to compare the methods of Exercises 4 and 5 with the Runge–Kutta methods of Exercise 3. What are your estimates of the order of the truncation errors for the two predictor–corrector methods?

7　Derive the second-order Adams–Bashforth method as a quadrature formula. That is, find the quadrature rule

$$\int_0^1 F(t)dt \approx \beta_1 F(0) + \beta_2 F(-1)$$

which is exact for $F(t) = 1, t$.

8　Derive the fourth-order Adams–Moulton formula AM4. (This will use nodes 1, 0, -1, -2 and be exact for cubic polynomials, see (6.25).)

9　Derive the fourth-order Adams–Bashforth formula. (This will use nodes 0, -1, -2, -3 and be exact for cubic polynomials.)

10　Apply the third- and fourth-order Adams–Bashforth methods AB3 and AB4 to the standard problem using $N = 20$ steps on $[0, 4]$. Compare the results with those of the fourth-order Runge–Kutta method RK4 with $N = 10$.

11　Repeat Exercise 10, doubling the numbers of steps used twice ($N = 40$ and $N = 80$ for the Adams–Bashforth methods). What are your estimates of the orders of their truncation errors?

12　Use the predictor–corrector methods with AB3 and AB4 as predictors and AM4 as corrector. Use $N = 20, 40, 80$. Compare the results with those for RK4 using $N = 10, 20, 40$.

6.4　Systems of differential equations

In this short section, we consider solving a system of first-order initial-value problems. Such a system can be written in vector form as

$$\mathbf{y}' = \mathbf{f}(x, \mathbf{y}), \quad \mathbf{y}(x_0) = \mathbf{y}_0 \tag{6.26}$$

Here $\mathbf{y}$ is an n-dimensional vector function of x, so that $\mathbf{y} : \Re \to \Re^n$ and $\mathbf{f}$ is an n-dimensional vector-valued function of $n + 1$ variables, $\mathbf{f} : \Re^{n+1} \to \Re^n$. (As is typical in discussing differential equations we are making no distinction here between a *function* and its *values*. The meaning should always be clear from the context.) In component form, (6.26) is

$$
\begin{aligned}
y_1' &= f_1(x, y_1, y_2, \ldots, y_n), & y_1(x_0) &= y_{1,0} \\
y_2' &= f_2(x, y_1, y_2, \ldots, y_n), & y_2(x_0) &= y_{2,0} \\
\cdots \quad &\cdots \quad \cdots \\
y_n' &= f_n(x, y_1, y_2, \ldots, y_n), & y_n(x_0) &= y_{n,0}
\end{aligned}
\tag{6.27}
$$

The techniques available for initial-value problems of this sort are essentially the same as those for single differential equations that we have already discussed. The key difference is that, at each step, a *vector* step must be taken. MATLAB's easy handling of matrices and vectors makes this a minor complication as we shall see. (For 'hand-calculation' this becomes tedious at best, and so we shall consider only computer solutions.)

Since the classical fourth-order Runge–Kutta method RK4 is both easy to program and very efficient, we shall concentrate on its use for solving systems. The MATLAB code below is almost identical to that in Section 6.2. The only difference is the use of vector quantities for the initial conditions and in the main loop.

Program

RK4 for an initial-value system

```
function sol=RK4sys(fcn,a,b,y0,N)
%Generates RK4 solution to y'=f(x,y) on [a,b]
%for a system of differential equations using N steps
%Vector initial condition is y(a)=y0
%y0 is a row vector
h=(b-a)/N;
x=a+(0:N)*h;
y(1,:)=y0;
for k=1:N
    k1=feval(fcn,x(k),y(k,:));
    k2=feval(fcn,x(k)+h/2,y(k,:)+h*k1/2);
    k3=feval(fcn,x(k)+h/2,y(k,:)+h*k2/2);
    k4=feval(fcn,x(k)+h,y(k,:)+h*k3);
    y(k+1,:)=y(k,:)+h*(k1+2*(k2+k3)+k4)/6;
end
sol=[x',y];
```

Example 9   Predator–Prey model: let $B(t), A(t)$ represent the populations of a bacteria and its antibodies at time t; a simple predator–prey model for this system has the form

$$B'(t) = \alpha B(t) - \beta B(t)A(t)$$
$$A'(t) = -\gamma A(t) + \delta A(t)B(t)$$

Here α represents the rate at which the bacterial population would grow in the absence of antibodies, γ is the rate at which the antibody population would decline in the absence of the bacteria. The 2 nonlinear terms represent the interaction of these two populations. (All the coefficients are assumed positive, with the attached signs implying the 'direction' of their

influence.) With *t* measured in days, find the populations of bacteria and antibodies after 4 days given initial populations

$$B(0) = 500, \quad A(0) = 200$$

with $\alpha = 0.7, \gamma = 0.2, \beta = 0.005, \delta = 0.001$. (The first 2 of these imply that the unchecked bacteria population would more than double in 1 day, and the unchallenged antibody population declines about 20% per day.)

The right-hand side of this system is represented by the m-file

```
function r=predprey(t,y)
r(1)=0.7*y(1)-0.005*y(1)*y(2);
r(2)=-0.2*y(2)+0.001*y(1)*y(2);
```

where y(1) represents the bacteria population B, and y(2) represents the antibodies A.

The solution is then computed by the commands

```
» BA0=[500,200];
» ba=rk4sys('predprey',0,4,BA0,100);
```

A good representation of the solution is obtained by plotting the bacterial population against the antibody population at the corresponding time:

```
» plot(ba(:,3),ba(:,2))
```

The resulting plot is shown in Figure 6.6. It is evident that the bacteria population declines steadily while the antibodies multiply until their population peaks at about 260. At this time (just under 2 days) the antibodies appear to have 'won' and both populations decrease.

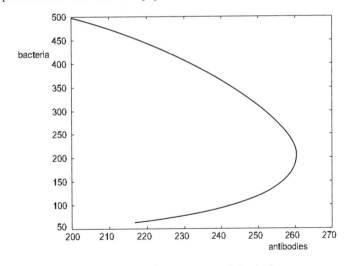

Figure 6.6 Predator–prey model solution

Another important class of differential equations, higher-order equations, can easily be rearranged as systems of first-order equations. They can then be solved in this same manner. We shall illustrate the principle for a second-order initial-value problem which has the general form

$$y'' = f(x, y, y'); \quad y(x_0) = y_0, \ y'(x_0) = y'_0 \tag{6.28}$$

Define two new variables (functions) by

$$u_1 = y, \quad u_2 = y'$$

Then, by definition, we have

$$u'_1 = u_2$$

and the differential equation (6.28) can be rewritten as

$$u'_2 = f(x, u_1, u_2)$$

The initial-value problem therefore becomes the first-order system

$$
\begin{aligned}
u'_1 &= u_2 & u_1(x_0) &= y_0 \\
u'_2 &= f(x, u_1, u_2) & u_2(x_0) &= y'_0
\end{aligned}
$$

which could now be solved in just the same way as was used for the predator–prey model in Example 9.

Example 10   Solve the initial-value problem

$$y'' = 4xy' + 2(1 - 2x^2)y; \quad y(0) = 0, y'(0) = 1$$

on the interval $[0, 2]$ using 100 steps of RK4

First we rearrange the equation as a system:

$$
\begin{aligned}
u'_1 &= u_2 & u_1(0) &= 0 \\
u'_2 &= 4xu_2 + 2(1 - 2x^2)u_1 & u_2(0) &= 1
\end{aligned}
$$

This system can now be solved using the same approach as above. The true solution to this equation is $y = x \exp(x^2)$. The computed solution and the error curve are plotted in Figure 6.7.

It is plain that the magnitude of the error is growing steadily as we expect for these methods. It is also clear that the error remains very small (less than 3.5×10^{-8}) throughout this range even with a relatively small number of steps.

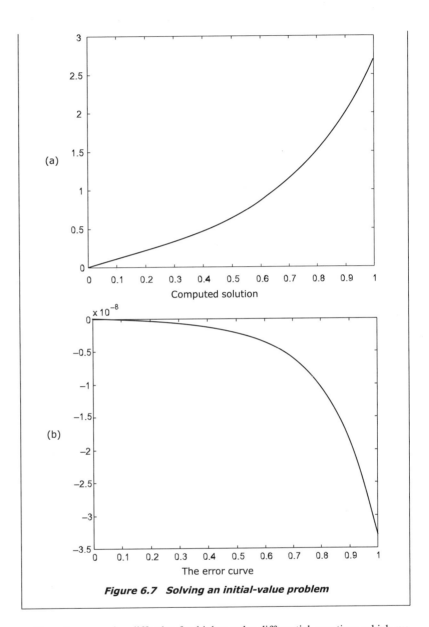

Figure 6.7 Solving an initial-value problem

There is one major difficulty for higher-order differential equations which we have not yet faced. The examples used have been *initial-value* problems. For higher-order differential equations it is common to be given boundary conditions at two distinct points. These boundary-value problems are the subject of the remaining sections of this chapter.

**Exercises:
Section 6.4**

1   Solve the initial-value system $y_1' = x + y_1 - 2y_2$, $y_2' = 2y_1 + y_2$ with $\mathbf{y}(0) = (0,0)$ over the interval $[0,4]$ using 200 steps of the fourth-order Runge–Kutta method. Plot the functions y_1, y_2 on the same axes.

2   Solve the same predator–prey equations as in Example 9 over longer time spans $[0,10]$ and $[0,20]$ using $h = 0.1$ in both cases.

3   Vary the starting conditions for the predator–prey problem of Exercise 2. Fix $B(0) = 500$ and use $A(0) = 100, 200, 300$ over the intervals $[0,20]$, $[0,20]$ and $[0,30]$ respectively. Plot the solutions on the same axes.

4   Solve the Bessel equation $x^2 y'' + xy' + (x^2 - 1)y = 0$ with initial conditions $y(1) = 1$, $y'(1) = 0$ over the interval $[1,15]$ using $N = 280$ steps. Plot the computed solution.

5   Solve the second-order nonlinear equation $y'' = 2yy'$ over the interval $[0,1.5]$ using $N = 150$ steps for the following initial conditions:

(a) $(y(0), y'(0)) = (0,1)$ (b) $(y(0), y'(0)) = (0,2)$

(c) $(y(0), y'(0)) = (1,1)$ (d) $(y(0), y'(0)) = (1,-1)$.

Plot all the solutions on the same axes.

6   A projectile is fired with initial speed 1000 m/s at angle α^0 from the horizontal. A good model for the path uses an air-resistance force proportional to the square of the speed. This leads to the equations

$$x'' + cx'\sqrt{x'^2 + y'^2} = 0, \quad y'' + cy'\sqrt{x'^2 + y'^2} = -g$$

Take the constant $c = 0.005$. The initial conditions are $x(0) = y(0) = 0$ and $(x'(0), y'(0)) = 1500(\cos\alpha, \sin\alpha)$. Convert this system of equations to a set of 4 first-order equations and solve them over the time-interval $[0,20]$ for each of the launch angles $\alpha = 10 : 10 : 80$. Plot the trajectories on the same set of axes.

6.5   Boundary-value problems: (1) shooting methods

The basic idea of *shooting methods* for the solution of a 2-point boundary-value problem is that we embed the solution to a related initial-value problem in an equation-solver. The resulting equation is then solved so that the final solution to the initial-value problem also satisfies the boundary conditions. The specific details will, of course, vary according to the particular type of equation and the nature of the boundary conditions.

We shall concentrate on the solution of a second-order differential equation with boundary conditions which specify the values of the solution at 2 distinct points. That is, we wish to solve

$$y'' = f(x, y, y') \tag{6.29}$$

subject to the *boundary conditions*

$$y(a) = y_a, \quad y(b) = y_b \tag{6.30}$$

Under reasonable conditions, we expect the initial-value problem

$$y'' = f(x, y, y'); \quad y(a) = y_a, y'(a) = z \tag{6.31}$$

to have a solution for each value of the *unknown* parameter z. Our objective is to find the appropriate value of z so that the solution of (6.31) hits the 'target' value $y(b) = y_b$.

How do we convert this into an equation for z?

Denote the solution of (6.31) for any value of z by $y(x; z)$ and define the function $F(z)$ by

$$F(z) = y(b; z) - y_b \tag{6.32}$$

We want to solve the equation

$$F(z) = 0 \tag{6.33}$$

for then

$$y(b; z) = y_b$$

and so the function $y(b; z)$ is a solution the differential equation (6.29) which satisfies the boundary conditions (6.30).

How do we solve (6.33) for z?

One relatively straightforward approach is to use the secant method. We select 2 initial guesses (or estimates) z_0, z_1 for z and solve the initial-value problem (6.31) for each of them. These yield $F(z_0)$ and $F(z_1)$ after which we can apply the secant method (Section 2.5) to generate our next estimate z_2. The secant iteration can then proceed as usual – the only difference being the need to solve a second-order initial-value problem on each iteration.

This approach lends weight to our search for efficient techniques for solving some of the fundamental problems of scientific computing. The need for an efficient equation-solver is much easier to appreciate when evaluating the 'function' entails the accurate solution of a system of differential equations. Similarly, if the differential equation may need to be solved repeatedly for different initial conditions, the need for efficient methods of solving differential equations becomes apparent.

We illustrate the process using examples. The first iteration is described in considerable detail. Later we show how to set up the function F as a MATLAB m-file so that the secant method can be used just as if it were a conventional function defined by some algebraic expression.

Example 11　**Solve the Bessel equation $x^2 y'' + xy' + (x^2 - 1)y = 0$ with boundary conditions $y(1) = 1,\ y(15) = 0$ by the shooting method**

First, we saw, in Exercise 4 of Section 6.4, the solution of the initial-value problem $x^2 y'' + xy' + (x^2 - 1)y = 0$ with $y(1) = 1,\ y'(1) = 0$ and that it yields a positive value for $y(15)$. This corresponds to $F(0)$ since the initial value is $y'(1) = 0$ and, in this case, $F(z) = y(15, z) - 0$. Specifically with $N = 280$ steps, we can use

```
» solz=rk4sys('bess1',1,15,[1,0],280);
» Fz=solz(281,2)
Fz =
    0.2694
```

which is to say $F(0) = 0.2694$.

For a second value we shall take $z = 1$:

```
» solz=rk4sys('bess1',1,15,[1,1],280);
» Fz=solz(281,2)
Fz =
    0.5356
```

which happens to be a worse guess than the first one.

Next, a secant iteration is used:

$$z_2 = z_1 - \frac{z_1 - z_0}{F(z_1) - F(z_0)} F(z_1)$$

$$= 1 - \left(\frac{1 - 0}{0.5356 - 0.2694}\right) 0.5356$$

$$= -1.012021$$

The next iteration begins with the solution of the differential equation with initial conditions $y(1) = 1, y'(1) = -1.012021$:

```
» solz=rk4sys('bess1',1,15,[1,}- 1.012021],280);
» Fz=solz(281,2)
Fz =
    -9.5985e-005
```

We have found the desired solution to reasonable accuracy using just 1 iteration of the secant method. It is this last solution of the differential equation. The two initial guesses and the final solution are plotted together in Figure 6.8 along with the target point. It is apparent that we have scored a 'bull's-eye'.

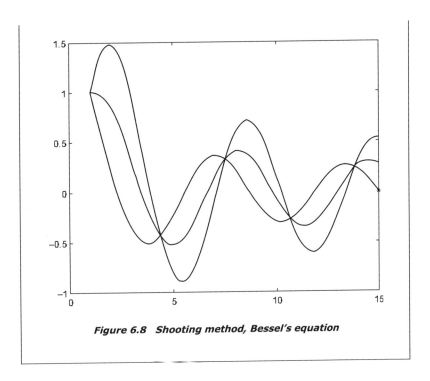

Figure 6.8 Shooting method, Bessel's equation

Why is this method so successful in this case? Can we always expect such good results?

Unfortunately the answer to the second question is 'No'. The situation here is special. Bessel's equation is a linear differential equation even though its *coefficients* are nonlinear functions of the independent variable. The general solution of such a differential equation is therefore a linear combination of two linearly independent solutions. (In this particular case, these are usually written as the Bessel functions of the first and second kinds J_1 and Y_1.) It follows that $y(15)$ is just a linear combination of $J_1(15)$ and $Y_1(15)$. Details of Bessel functions are not important to this argument – what is important is that the equation $F(z) = 0$ is then a *linear* equation. Since the secant method is based on linear interpolation, it will find the solution of such an equation in a single iteration.

Thus the shooting method using the secant algorithm as the equation-solver will solve a linear 2-point boundary problem in just 1 iteration. In general we cannot anticipate such good performance. In Example 12, we consider a nonlinear differential equation and develop a more complete algorithm for shooting methods.

Example 12 **Solve the boundary-value problem**

$$y'' = 2yy'; \quad y(0) = 1, y(1) = -1$$

using the shooting method

We take as initial guesses $y'(0) = 0$ which yields the solution $y(x) = 1$, and $y'(0) = -1$.

The following MATLAB commands yield the solutions of the initial-value problems and, from these, the amount by which we miss the target:

```
» z=0; y0=[1,z];
» sz=rk4sys('trig1',0,1,y0,200);
» Fz=sz(201,2)+1
Fz =
    2
» z=-1;
» y0=[1,z];
» sz=rk4sys('trig1',0,1,y0,200);
» Fz=sz(201,2)+1
Fz =
    0.3105
```

Note that the value of Fz is given by the value of the solution at the final point (corresponding to $x = 1$) plus 1 since, for this example, we have

$$F(z) = y(1;z) - (-1)$$

The solution curves for these initial guesses are plotted in Figure 6.9.

To complete the first secant iteration, we set

```
» z=z1-(z1-z2)/(Fz1-Fz2)*Fz1
z =
    -1.1838
```

where we have used z2=0, z1=-1. The solution of the initial-value problem corresponding to this initial slope is then computed. It, too, is plotted in Figure 6.9 and can be seen to get significantly closer to the 'target' without yet being close enough.

Clearly, we need to automate this iteration. The key step is to create a function which generates $F(z)$ for any initial slope $y'(0) = z$. The steps used above show how to achieve this. For this example the following m-file can be used:

```
function Fz=trigshoot(z)
% Function for shooting method for y'=2yy'
% 'trig1' implements this differential equation as a system
y0=[1,z];
sz=rk4sys('trig1',0,1,y0,200);
Fz=sz(201,2)+1;
```

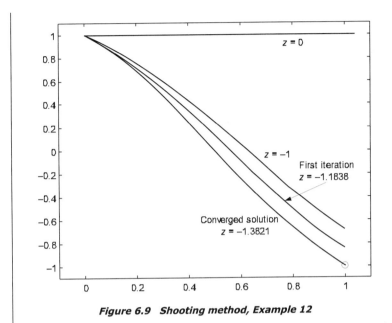

Figure 6.9 Shooting method, Example 12

The function trig1.m referred to here is just

```
function r=trig1(x,y)
r(1)=y(2);
r(2)=2*y(1).*y(2);
```

At this stage the regular secant method can be applied to the equation $F(z) = 0$ using the command

```
» zsol=secant('trigshoot',0,-1,1e-3)
zsol =
   -1.3821
```

The solution for $y'(0) = -1.3821$ is also plotted in Figure 6.9. It is clear that it hits the target. The actual error at the right-hand end of the interval was about 5×10^{-9}.

It is clear that for examples such as this, where the dependence on the initial conditions is not too sensitive, the shooting method provides an efficient method of solving 2-point boundary-value problems. It should also be observed that the boundary conditions could be specified in a number of ways. The basic approach to the solution remains unchanged. For example, if we had specified the *slope* of the solution at $x = 1$ in Example 12 we would simply have compared the computed slope with this value. (See the Exercises.)

Complications can arise with sensitive dependence on initial conditions. Another source of potential difficulty is encountered in the type of problem that gave shooting methods their name.

Suppose we wish to compute the launch angle for a projectile to strike a specific target. As we saw in the previous section, for the second-order air-resistance model the differential equations are a pair of nonlinear second-order equations which we can rewrite as a system of 4 equations. The major difference is that the boundary conditions specify a *position* whereas the independent variable is *time*. That is, the boundary conditions are given at an unspecified value of the independent variable.

Such a problem can still be solved using the shooting method. In essence, we examine the solution to check the y-coordinate at the point where the x-coordinate is closest to the desired target and use this to measure how far we are off-target. (This is explored in the Exercises.)

Exercises:
Section 6.5

1   Solve the Bessel equation $x^2 y'' + xy' + (x^2 - 4)y = 0$ with the boundary conditions $y(1) = 1$, $y(10) = 1$.

2   Solve the Bessel equation $x^2 y'' + xy' + (x^2 - 4)y = 0$ with the boundary conditions $y(1) = 10$, $y'(10) = -1$.(This is similar to Exercise 1 except that the right-hand boundary condition is given in terms of the slope. Note that the other component of the solution to the system gives the values of the slope.)

3   Solve the equation $y'' = 2yy'$ with boundary conditions (a) $y(0) = 0$, $y(1) = 1$, (b) $y(0) = -1$, $y'(1) = 1$, (c) $y'(0) = 1$, $y(1) = 0$

4   A projectile is launched from the origin with initial speed 200 m/s. Find the appropriate launch angle so that the projectile hits its target which is at coordinates (100,30). Assume the air resistance is proportional to the square of the speed with the constant 0.01 so that the differential equations are

$$x'' + 0.01 x' \sqrt{x'^2 + y'^2} = 0, \quad y'' + 0.01 y' \sqrt{x'^2 + y'^2} = -g$$

For a given launch angle, compute the trajectory over a period of 10 seconds using a time-step of 0.1 seconds. Find the x-value in your solution that is closest to 100. For $h(\alpha)$ take the corresponding y-value and solve $h(\alpha) - 30 = 0$ by the secant method.

6.6 Boundary-value problems: (2) finite difference methods

In this short section, we consider an alternative to shooting methods for the special situation where the differential equation is linear. Specifically, we consider a 2-point boundary problem of the form

$$y'' + a(x)y' + b(x)y = f(x); \quad y(x_0) = y_0, y(x_N) = y_N \qquad (6.34)$$

where $[x_0, x_N]$ is the interval over which the solution is sought and N is the number of steps to be used.

We need to compute the values $y_1, y_2, \ldots, y_{N-1}$ of the solution at the intermediate points

$$x_k = x_0 + kh \quad (k = 1, 2, \ldots, N - 1)$$

where the steplength h is given by

$$h = \frac{x_N - x_0}{N}$$

The basic idea behind finite difference methods is that we use difference approximations for the derivatives to obtain a system of linear equations for the y-values.

We shall use our second-order difference formulas from Section 5.6:

$$y'(x_k) \approx \frac{y(x_{k+1}) - y(x_{k-1})}{2h} \tag{6.35}$$

$$y''(x_k) \approx \frac{y(x_{k+1}) - 2y(x_k) + y(x_{k-1})}{h^2} \tag{6.36}$$

Throughout the remainder of this section, we shall use the notation

$$a(x_k) = a_k,$$

$$b_k = b(x_k),$$

$$f_k = f(x_k)$$

and use y_k for our approximate value of $y(x_k)$.

At any point x_k $(1 \le k < N)$, using the approximations (6.35) and (6.36), the differential equation in (6.34) becomes

$$\frac{y_{k+1} - 2y_k + y_{k-1}}{h^2} + a_k \frac{y_{k+1} - y_{k-1}}{2h} + b_k y_k = f_k$$

Multiplying by $2h^2$ and collecting terms this yields

$$(2 - ha_k)y_{k-1} + (2h^2 b_k - 4)y_k + (2 + ha_k)y_{k+1} = 2h^2 f_k \tag{6.37}$$

which provides a *tridiagonal* system of linear equations for the unknown values $y_1, y_2, \ldots, y_{N-1}$. The first and last of these equations $(k = 1, N - 1)$ must be adjusted to absorb the terms relating to y_0, y_N into their respective right-hand sides.

The resulting system can be written as

$$
\begin{bmatrix}
2h^2b_1 - 4 & 2 + ha_1 \\
2 - ha_2 & 2h^2b_2 - 4 & 2 + ha_2 \\
& 2 - ha_3 & 2h^2b_3 - 4 & 2 + ha_3 \\
& & \ddots & \ddots & \ddots \\
& & & 2 - ha_{N-2} & 2h^2b_{N-2} - 4 & 2 + ha_{N-2} \\
& & & & 2 - ha_{N-1} & 2h^2b_{N-1} - 4
\end{bmatrix}
\begin{bmatrix}
y_1 \\ y_2 \\ y_3 \\ \vdots \\ y_{N-2} \\ y_{N-1}
\end{bmatrix}
$$

$$
=
\begin{bmatrix}
2h^2f_1 - (2 - ha_2)y_0 \\
2h^2f_2 \\
2h^2f_3 \\
\\
2h^2f_{N-2} \\
2h^2f_{N-1} - (2 + ha_{N-1})y_N
\end{bmatrix}
\tag{6.38}
$$

Recall that we also encountered tridiagonal systems of linear equations in computing the coefficients for cubic spline interpolation in Chapter 4. Tridiagonal systems are important special cases of linear systems of equations, the subject of Chapter 7, the next, and final, chapter. In Example 13, we again use MATLAB's linear system-solver as a 'black-box' to solve this aspect of the problem.

Example 13   **Solve the Bessel equation $x^2y'' + xy' + (x^2 - 1)y = 0$ subject to $y(1) = 1$, $y(15) = 0$ using the finite difference method just described with $N = 280$ steps**

First we must rewrite the differential equation in the form

$$
y'' + \frac{1}{x}y' + \left(1 - \frac{1}{x^2}\right)y = 0
$$

so that $a(x) = 1/x$, $b(x) = 1 - 1/x^2$ and $f(x) = 0$. Also $h = (15 - 1)/280 = 1/20$ so that $x_k = 1 + k/20$.

The following MATLAB commands generate the matrix of coefficients in (6.38) for this example, and then the standard solver is used to generate the solution and the plot in Figure 6.10.

```
»  h=1/20;
»  N=280;
»  x=1+(0:N)*h;
»  a=1./x;
»  b=1−1./x.^2;
»  f=zeros(N−1,1);
»  M=diag(2*h^2*b(2:N−1) −4);
```

```
»  for k=1:N−2
M(k,k+1)=2+h*a(k+1);
M(k+1,k)=2}−h*a(k+2);
end
»  y(1)=1; y(N+1)=0;
»  f(1)=f(1)−(2− h*a(2))*y(1);
»  f(N−1)=f(N− 1)−(2+h*a(N))*y(N+1);
»  y(2:N)=M\f;
»  plot(x,y)
```

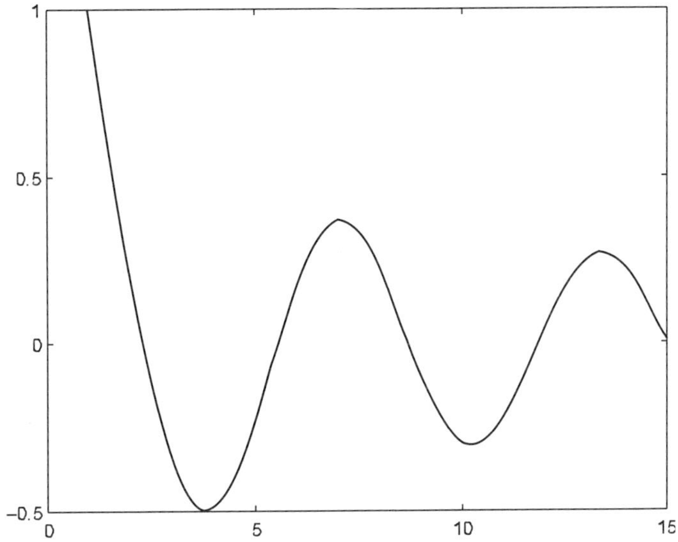

Figure 6.10   Finite difference solution, Example 13

Note that the matrix used in this example would be too large for some of the earlier versions of the *Student Edition* of MATLAB. Using significantly fewer points would be possible but that would obviously degrade the accuracy of the solution. This difficulty arises *only* because we are using the 'black box' linear equation-solver.

**Exercises:
Section 6.6**

1   Solve the Bessel equation $x^2y'' + xy' + (x^2 - 1)y = 0$ subject to $y(1) = 1$, $y(10) = 1$ by the second-order finite difference method using $N = 90$ steps.

2   Repeat Exercise 1 using 30 steps and 60 steps. Plot all the computed solutions on the same axes.

3   Solve $x^2y'' + xy' + (x^2 - 4)y = 0$ subject to $y(1) = 1, y(10) = 2$ by the second-order finite difference method with steplength $h = 0.2$.

4   Solve $y'' + 2xy + 2y = 3e^{-x} - 2xe^{-x}$ subject to $y(-1) = e + 3/e, y(1) = 4/e$ using the second-order finite difference method with $N = 10, 20, 40$ steps. Plot these solutions and the true solution $y = e^{-x} + 3e^{-x^2}$. Also, on a separate set of axes, plot the 3 error curves. Estimate the order of the truncation error for this method.

6.7   MATLAB functions for ordinary differential equations

MATLAB has a number of functions for solving ordinary differential equations and for displaying the results of these.

ode23   is an adaptive Runge–Kutta–Fehlberg method using a pair of Runge–Kutta formulas, one second-order and one third-order. This routine is intended to provide a moderate level of accuracy for differential equations which are not 'stiff'.

ode45   is a more accurate version of ode23 using a fourth-order and fifth-order pair of Runge–Kutta formulas.

The syntax for the two is essentially the same. For example, to solve Bessel's equation $x^2y'' + xy' + (x^2 - 1)y = 0$ subject to the initial conditions $y(1) = 2$, $y'(1) = 1$ over the interval $[1, 8]$ we could use

[x,y]=ode23('bess1',[1,8],[2;1]);

where, as usual, bess1 is the function m-file for the right-hand side of the corresponding *system* of first-order differential equations. Note that the domain for our solution is specified as a *row* vector while the initial conditions **must** be given as a *column* vector. The output of the function bess1 must also be arranged as a column vector.

The output [x,y] consists of a column of values of the independent variable x and a matrix y whose columns are the components of the solution. In this case the first column has the values of y and the second those of y'.

There are optional arguments to these functions to adjust the tolerance used in the solution and to force intermediate results to be displayed.

The command

» plot(x,y)

generates the plot in Figure 6.11, of both y and y' against x.

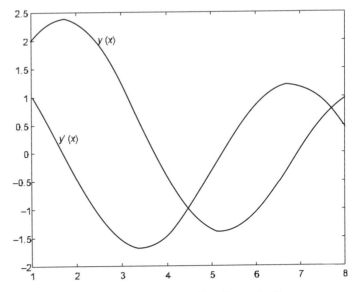

Figure 6.11   Plot resulting from ode23

Other functions   MATLAB has several other differential equation-solvers. ode113 is another 'nonstiff' solver based on using an Adams–Bashforth–Moulton predictor–corrector method. It can be used as a variable-order method which can deliver greater accuracy than ode45.

The functions used for 'stiff' differential equations are ode15s and ode23s. Since we have not defined what is meant by a stiff differential equation, we do not cover their details here. The syntax is very similar to that for ode23 above. Full details of these can be obtained from the MATLAB documentation.

7 Linear Equations

Aims and objectives

In previous chapters we saw several instances where we need to solve a system of linear equations. We begin with the basic Gauss elimination process, and its practical implementation. We see modifications that are both necessary and powerful. One important objective is therefore to answer a recurring problem. A further application, to least squares approximation, or linear regression, completes the loop back to function approximation. Finally, we discuss basic methods for finding *eigenvalues*. The objective here is only to introduce, in a rudimentary way, an area of great practical importance in more advanced applications, which remains a major area of theoretical and practical research. (In fact, almost every topic discussed in this book is an active research subject.)

7.1 Introduction

We have observed, in several situations, the need to solve a system of linear simultaneous equations. This arose, for example, in the study of interpolation, whether by polynomials or by splines, in quadrature and again in finite difference methods for boundary-value problems. In the cases of cubic spline interpolation and finite difference methods, we saw the need to handle an important special case where the matrix of coefficients is tridiagonal. Linear systems also arise naturally in direct application and in the solution of partial differential equations. As an application later we shall also see that one common approach to approximation of a function also reduces to a system of linear equations.

In the case where the number of equations and the number of unknowns are both small, such systems can be readily solved by hand using the familiar idea of eliminating the unknowns one by one by solving 1 equation for one of them in terms of the others. Eventually, we are left with just 1 equation in 1 unknown. Its solution can then be substituted back into the previous equation to find the next unknown, and so on. This is a terse and somewhat imprecise description of the process of *Gauss elimination* which we discuss more systematically in the next section.

There are direct methods available such as Cramer's rule which for a system of N equations in N unknowns requires the evaluation of $(N + 1)$ $N \times N$ determinants. Even for a 10×10 system this requires some 359 million multiplications. This is certainly not a task to be undertaken by hand! Nor is it a sensible way of approaching the problem on a computer. For a 20×20 system the number of multiplications alone would become 21! which is approximately 5×10^{19}. On a very fast (1 *gigaflop*) computer, these multiplications would take 5×10^{10} seconds, or approximately 1600 *years*. (One gigaflop is one billion, 10^9, floating-point operations per second.) Our solution would not be computed until long after the Y3K problem has been fixed!

Note that 20×20 systems are not large. In Chapter 6 we solved a 279×279 linear system in order to solve Bessel's equation. Obviously Cramer's rule was not employed! In many practical problems systems with several *thousand* equations and unknowns are common.

We shall concentrate our efforts in this brief introduction to an enormous subject, on the solution of a square system of equations with a single right-hand side. That is, we shall solve systems of the form

$$A\mathbf{x} = \mathbf{b} \tag{7.1}$$

where $\mathbf{x}, \mathbf{b}$ are n-vectors and A is an $n \times n$ real matrix. In full, this system would be written as

$$
\begin{bmatrix}
a_{11} & a_{12} & a_{13} & \cdots & a_{1n} \\
a_{21} & a_{22} & a_{23} & \cdots & a_{2n} \\
a_{31} & a_{32} & a_{33} & \cdots & a_{3n} \\
\cdot & \cdot & \cdot & \cdot & \cdot \\
a_{n1} & a_{n2} & a_{n3} & \cdots & a_{nn}
\end{bmatrix}
\begin{bmatrix}
x_1 \\ x_2 \\ x_3 \\ \vdots \\ x_n
\end{bmatrix}
=
\begin{bmatrix}
b_1 \\ b_2 \\ b_3 \\ \vdots \\ b_n
\end{bmatrix}
\tag{7.2}
$$

or even, writing each equation fully,

$$
\begin{aligned}
a_{11}x_1 + a_{12}x_2 + a_{13}x_3 + \cdots + a_{1n}x_n &= b_1 \\
a_{21}x_1 + a_{22}x_2 + a_{23}x_3 + \cdots + a_{2n}x_n &= b_2 \\
a_{31}x_1 + a_{32}x_2 + a_{33}x_3 + \cdots + a_{3n}x_n &= b_3 \\
\cdots \quad \cdots \quad \cdots \quad \cdots \quad \cdots &= \cdots \\
a_{n1}x_1 + a_{n2}x_2 + a_{n3}x_3 + \cdots + a_{nn}x_n &= b_n
\end{aligned}
\tag{7.3}
$$

There are many situations of interest which can be treated by extensions of the ideas presented here. The case of multiple right-hand sides can be handled using the same methods as are presented in following sections with very little modification. The essential change is the obvious one: whatever operations are performed on the one right-hand side must be applied to them all. We shall discuss this extension very briefly later.

Beyond the question of solving systems of linear equations there is a second fundamental problem of linear algebra, the *eigenvalue* problem, which requires the solution of the equation

$$A\mathbf{x} = \lambda\mathbf{x} \qquad (7.4)$$

for the *eigenvalues* λ and their associated *eigenvectors* $\mathbf{x}$. There are many techniques available for this problem, most of which are beyond our present aims. However, we shall introduce one basic method and an extension which makes use of our linear equations techniques.

7.2 Gauss elimination

We begin with the basic ideas of Gauss elimination. Later in the section, we shall consider potential difficulties that arise in the basic method. We also consider in some detail the special case of tridiagonal systems of equations that we have encountered in a couple of places already.

The first step in the Gauss elimination process is to eliminate the unknown x_1 from every equation in (7.3) except the first. The way this is achieved is to subtract from each subsequent equation an appropriate multiple of the first equation. Since the coefficient of x_1 in the first equation is a_{11} and that in the j-th equation is a_{j1}, it follows that subtracting $m = \dfrac{a_{j1}}{a_{11}}$ times the first equation from the j-th one will result in the coefficient of x_1 becoming

$$a_{j1} - ma_{11} = a_{j1} - \frac{a_{j1}}{a_{11}}a_{11} = 0$$

The resulting system of equations is then

$$
\begin{aligned}
a_{11}x_1 + a_{12}x_2 + a_{13}x_3 + \cdots + a_{1n}x_n &= b_1 \\
a'_{22}x_2 + a'_{23}x_3 + \cdots + a'_{2n}x_n &= b'_2 \\
a'_{32}x_2 + a'_{33}x_3 + \cdots + a'_{3n}x_n &= b'_3 \\
\cdots \quad\cdots\quad\cdots\quad\cdots\ \ &= \cdots \\
a'_{n2}x_2 + a'_{n3}x_3 + \cdots + a'_{nn}x_n &= b'_n
\end{aligned}
\qquad (7.5)
$$

or, in matrix notation,

$$
\begin{bmatrix}
a_{11} & a_{12} & a_{13} & \cdots & a_{1n} \\
0 & a'_{22} & a'_{23} & \cdots & a'_{2n} \\
0 & a'_{32} & a'_{33} & \cdots & a'_{3n} \\
\vdots & & & & \vdots \\
0 & a'_{n2} & a'_{n3} & \cdots & a'_{nn}
\end{bmatrix}
\begin{bmatrix}
x_1 \\ x_2 \\ x_3 \\ \vdots \\ x_n
\end{bmatrix}
=
\begin{bmatrix}
b_1 \\ b'_2 \\ b'_3 \\ \vdots \\ b'_n
\end{bmatrix}
\qquad (7.6)
$$

where the modified coefficients are given by

$$a'_{jk} = a_{jk} - \frac{a_{j1}}{a_{11}} a_{1k}$$

$$b'_j = b_j - \frac{a_{j1}}{a_{11}} b_1 \qquad (7.7)$$

This particular operation can be described by the loop

for j=2:n

$$m := \frac{a_{j1}}{a_{11}}; \; a_{j1} := 0$$

for k=2:n

$$a'_{jk} := a_{jk} - m a_{1k}$$

end

$$b'_j := b_j - m b_1$$

end

In the general algorithm, and in all that follows, the $'$ notation will be dropped and elements of the matrix and components of the right-hand side will just be overwritten by their new values. With this convention the modified rows of (7.6) represent the system

$$
\begin{bmatrix}
a_{22} & a_{23} & \cdots & a_{2n} \\
a_{32} & a_{33} & \cdots & a_{3n} \\
\cdot & \cdot & \cdot & \cdot \\
a_{n2} & a_{n3} & \cdots & a_{nn}
\end{bmatrix}
\begin{bmatrix}
x_2 \\
x_3 \\
\vdots \\
x_n
\end{bmatrix}
=
\begin{bmatrix}
b_2 \\
b_3 \\
\vdots \\
b_n
\end{bmatrix}
\qquad (7.8)
$$

The same idea can be applied to (7.8) to eliminate x_2 and so on until we are left with a *triangular* system

$$
\begin{bmatrix}
a_{11} & a_{12} & a_{13} & \cdots & a_{1n} \\
0 & a_{22} & a_{23} & \cdots & a_{2n} \\
0 & 0 & a_{33} & \cdots & a_{3n} \\
\vdots & \vdots & & \ddots & \vdots \\
0 & 0 & 0 & \cdots & a_{nn}
\end{bmatrix}
\begin{bmatrix}
x_1 \\
x_2 \\
x_3 \\
\vdots \\
x_n
\end{bmatrix}
=
\begin{bmatrix}
b_1 \\
b_2 \\
b_3 \\
\vdots \\
b_n
\end{bmatrix}
\qquad (7.9)
$$

This is the 'elimination' phase, or *forward elimination*, of Gauss elimination. It is easy to describe as a formal algorithm.

Algorithm 1   Forward elimination phase of Gauss elimination

Input $n \times n$ matrix A, right-hand side n-vector $\mathbf{b}$

Elimination

```
  for i=1:n-1
    for j=i+1:n
```
$$m := \frac{a_{ji}}{a_{ii}};\ a_{ji} := 0$$
```
      for k=i+1:n
```
$$a_{jk} := a_{jk} - m a_{ik}$$
```
      end
```
$$b_j := b_j - m b_i$$
```
    end
  end
```

Output   Triangular matrix A, modified right-hand side $\mathbf{b}$

Example 1   Use Gauss elimination to solve the system

$$\begin{bmatrix} 1 & 2 & 3 \\ 2 & 2 & 3 \\ 3 & 3 & 3 \end{bmatrix} \begin{bmatrix} x \\ y \\ z \end{bmatrix} = \begin{bmatrix} 6 \\ 7 \\ 9 \end{bmatrix}$$

The first step of the elimination is to subtract multiples of the first row from each of the other rows. The appropriate multipliers to use for the second and third rows are $2/1 = 2$ and $3/1 = 3$, respectively. The resulting system is

$$\begin{bmatrix} 1 & 2 & 3 \\ 0 & -2 & -3 \\ 0 & -3 & -6 \end{bmatrix} \begin{bmatrix} x \\ y \\ z \end{bmatrix} = \begin{bmatrix} 6 \\ -5 \\ -9 \end{bmatrix}$$

Next $(-3)/(-2) = 3/2$ of the second row must be subtracted from the third one in order to eliminate y from the third equation:

$$\begin{bmatrix} 1 & 2 & 3 \\ 0 & -2 & -3 \\ 0 & 0 & -3/2 \end{bmatrix} \begin{bmatrix} x \\ y \\ z \end{bmatrix} = \begin{bmatrix} 6 \\ -5 \\ -3/2 \end{bmatrix}$$

This completes the forward elimination for this example.

The final row now represents the equation

$$\frac{-3}{2}z = \frac{-3}{2}$$

from which we deduce that $z = 1$. Substituting this back into the second equation, we obtain

$$-2y - 3(1) = -5$$

and hence, $y = 1$. Finally substituting both these into the first equation, we get

$$x + 2(1) + 3(1) = 6$$

so that $x = 1$, also. The solution vector is therefore $[1, 1, 1]'$.

The second phase of the process used above is the *back substitution* phase of Gauss elimination. It can be summarized as follows.

Algorithm 2   Back substitution phase of Gauss elimination

Input $n \times n$ triangular matrix A, right-hand side n-vector $\mathbf{b}$
(typically the output from the forward elimination stage)
Elimination
$x_n := b_n/a_{nn}$ (Solve final equation)
   for i=n-1:-1:1
     for j=i+1:n
       $b_i := b_i - a_{ij}x_j$ (Substitute known values and subtract)
     end
     $x_i := b_i/a_{ii}$
   end
Output   Solution vector $\mathbf{x}$

Before proceeding any further, it is desirable to say a little more about the efficiency of Gauss elimination. For Cramer's rule, we observed that approximately $(n + 1)!$ multiplications would be needed to solve an $n \times n$ system. A careful counting of the numbers of multiplications and divisions in the two phases of the Gauss elimination algorithms reveals that approximately $n^3/3$ multiplications and $n^2/2$ divisions are needed. For the 20×20 system cited earlier this reduces the count from about 5×10^{19} down to about 2800. It is clear why we say *never* use Cramer's rule as a practical tool for solving linear systems!

The whole Gauss elimination process is very easy to program and provides a powerful basic tool for solving linear systems. *But* it has its shortcomings, as the following simple example demonstrates.

Example 2   Solve the following system using Gauss elimination

$$
\begin{bmatrix} 7 & -7 & 1 \\ -4 & 4 & -1 \\ 7 & 7 & -4 \end{bmatrix} \begin{bmatrix} x_1 \\ x_2 \\ x_3 \end{bmatrix} = \begin{bmatrix} 1 \\ -1 \\ 10 \end{bmatrix}
$$

Following the same algorithm as we used in Example 1, we begin by subtracting $(-4)/7$ times the first row from the second and $7/7$ times the first row from the third. The resulting *partitioned* matrix $\left[A \vdots \mathbf{b} \right]$ is

$$
\begin{bmatrix} 7 & -7 & 1 & \vdots & 1 \\ 0 & 0 & -3/7 & \vdots & -3/7 \\ 0 & 14 & -5 & \vdots & 9 \end{bmatrix}
$$

at which stage the algorithm breaks down since we next must subtract $14/0$ times the second row from the third!

Of course, the remedy in Example 2 is simple. By exchanging the second and third rows, we obtain the revised system

$$
\begin{bmatrix} 7 & -7 & 1 & \vdots & 1 \\ 0 & 14 & -5 & \vdots & 9 \\ 0 & 0 & -3/7 & \vdots & -3/7 \end{bmatrix}
$$

which is now a triangular system. The back substitution algorithm can now be used to complete the solution: $[x_1, x_2, x_3] = [1, 1, 1]$. (**Caution!** Not all linear systems have solution vectors consisting only of ones!)

Example 2 is an extreme example of the need for *pivoting*. Indeed that is precisely what we used to overcome the difficulty.

7.2.1 Pivoting in Gauss elimination

At each step of the basic Gauss elimination algorithm the diagonal entry that is being used to generate the multipliers is known as the pivot element. In Example 2, we saw the effect of a zero pivot. The basic idea of *partial pivoting* is to avoid zero (or near-zero) pivot elements by performing row interchanges. The details will be discussed shortly, but first we revisit Example 2 to demonstrate the need to avoid near-zero pivots.

Example 3   Solve the system of Example 2 using Gauss elimination with four decimal place arithmetic

At the first step, the multiplier used for the second row is $(-4)/7 = -0.5714$ to four decimals. The resulting system is then

$$\begin{bmatrix} 7 & -7 & 1 & \vdots & 1 \\ 0 & 0.0002 & -0.4286 & \vdots & -0.4286 \\ 0 & 14 & -5 & \vdots & 9 \end{bmatrix}$$

The multiplier used for the next step is then $14/0.0002 = 70000$ and subtracting $70000(-0.4286) = -30002$ from the final entries we get

$$\begin{bmatrix} 7 & -7 & 1 & \vdots & 1 \\ 0 & 0.0002 & -0.4286 & \vdots & -0.4286 \\ 0 & 0 & 29997 & \vdots & 30011 \end{bmatrix}$$

which gives the solutions (computed to four decimal places)

$$x_3 = 30011/29997 = 1.0005$$
$$x_2 = (-0.4286 - (-0.4286)(1.0005))/0.0002 = 1.0715$$
$$x_1 = (1 - (-7)(1.0715) - (1)(1.0005))/7 = 1.0714$$

4 decimal-place arithmetic has left the results accurate to only 1 decimal!

The problem in Example 3 is that the second pivot entry is dominated by its roundoff error (in this case, it *is* its roundoff error). This results in an unfortunately large multiplier being used which magnifies the roundoff errors in the other entries to the point where the accuracy of the final result is affected.

The basic idea of partial pivoting is to search the *pivot column* to find its largest element (in absolute value) on or below the diagonal. The *pivot row* is then interchanged with this row so that this largest element becomes the pivot for this step.

Example 4   Repeat Example 3 using partial pivoting

The first step is unchanged since the largest element in the first column is already on the diagonal. However, for the second step we would interchange rows 2 and 3 (since $14 > 0.0002$) to get

$$\begin{bmatrix} 7 & -7 & 1 & \vdots & 1 \\ 0 & 14 & -5 & \vdots & 9 \\ 0 & 0.0002 & -0.4286 & \vdots & -0.4286 \end{bmatrix}$$

The multiplier is now $0.0002/14 = 0$ to four decimal places. The final triangular system is then

$$\begin{bmatrix} 7 & -7 & 1 & \vdots & 1 \\ 0 & 14 & -5 & \vdots & 9 \\ 0 & 0 & -0.4286 & \vdots & -0.4286 \end{bmatrix}$$

which yields the exact solution $x_3 = x_2 = x_1 = 1$.

If 4 significant figures are used throughout, the accuracy of the solution is well-preserved. The multiplier for the last step becomes $0.0002/14 = 1.428 \times 10^{-5}$ from which we obtain the final equation:

$$-0.4285 x_3 = -0.4287$$

giving $x_3 = 0.4287/0.4285 = 1.000$ to four *significant* figures. Clearly the remaining components of the solution are unchanged.

The use of pivoting requires only a small change to the Gauss elimination algorithm.

Algorithm 3   Forward elimination for Gauss elimination with partial pivoting

Input $n \times n$ matrix A, right-hand side n-vector **b**

Elimination

    for i=1:n-1

        Find $p \geq i$ such that $|a_{pi}| = \max(|a_{ji}| : j \geq i)$

        Interchange rows i and p

        for j=i+1:n

            $m := \dfrac{a_{ji}}{a_{ii}}; \; a_{ji} := 0$

            for k=i+1:n

                $a_{jk} := a_{jk} - m a_{ik}$

            end

            $b_j := b_j - m b_i$

        end

    end

Output   Triangular matrix A, modified right-hand side **b**

MATLAB's max function makes this easy to implement. Remember though that the position of the maximum will be given relative to the vector to which max is applied. Thus, for example, to find the largest (absolute) entry in positions 6 through 10 of the vector $[1, 3, 5, 7, 9, 0, 2, 8, 6, 4]$ we can use

```
» v=[1,3,5,7,9,0,2,8,6,4];
  [mx,pmax]=max(v(6:10))
```

to get the output

```
mx =
  8
pmax =
  3
```

Here 8 is indeed the maximum entry in the appropriate part of the vector $[0, 2, 8, 6, 4]$ and it is the third component of this vector. To get its position in the original vector we simply add 5 to account for the 5 elements that were omitted from our search.

Program

MATLAB code for Gauss elimination with partial pivoting

```
function x=gepp(A,b)
% Gauss elimination with partial pivoting
% A is a square matrix, b is right-hand side vector
n=size(A,1); % n is number of rows of A
% Begin forward elimination
for i=1:n-1
    [m,p]=max(abs(A(i:n,i))); % find the pivot element and its position
    p=p+i-1;
    tmp=A(i,:); A(i,:)=A(p,:); A(p,:)=tmp; % row interchange
    tmpb=b(i); b(i)=b(p); b(p)=tmpb; % interchange on right-hand side
    for j=i+1:n
      m=A(j,i)/A(i,i);
      A(j,i)=0;
      for k=i+1:n
        A(j,k)=A(j,k)-m*A(i,k);
      end
      b(j)=b(j)-m*b(i);
    end
end   % End of forward elimination
% Begin back substitution
x(n)=b(n)/A(n,n);
for i=n-1:-1:1
    for j=i+1:n
      b(i)=b(i)-A(i,j)*x(j);
    end
    x(i)=b(i)/A(i,i);
end   % x is the solution vector
```

Note that it is not strictly necessary to perform the row interchanges explicitly. A permutation vector could be stored which carries the information as to the order in which the rows were used as pivot rows. For our current purpose, the advantage of such an approach is outweighed by the additional detail that is needed. MATLAB's vector operations make the explicit row interchanges easy and fast.

To test our algorithm, we shall use the *Hilbert matrix*, which is defined for dimension n as

$$H_n = \begin{bmatrix} 1 & 1/2 & 1/3 & \cdots & 1/n \\ 1/2 & 1/3 & 1/4 & \cdots & 1/(n+1) \\ 1/3 & 1/4 & 1/5 & \cdots & 1/(n+2) \\ . & . & . & \cdots & . \\ 1/n & 1/(n+1) & 1/(n+2) & \cdots & 1/(2n-1) \end{bmatrix}$$

The elements are

$$h_{ij} = \frac{1}{i+j-1}$$

For the right-hand side, we shall impose a simple test for which it is easy to recognize the correct solution. By taking the right-hand side as the vector of row-sums, we force the true solution to be a vector of ones. Note that MATLAB's sum function sums the *columns* of a matrix. Applying this to the transpose of a matrix will give a row vector consisting of the row-sums of the original matrix. Taking the transpose again yields the desired column vector.

Example 5   Solve the Hilbert matrix system of dimension 6 using Gauss elimination with partial pivoting

The n-dimensional Hilbert matrix is a built-in function hilb(n) in MATLAB. We can therefore use the following commands to generate the solution of ones:

```
» H=hilb(6);
» rsH=sum(H')';
» x=gepp(H,rsH);
» format short
» x'
ans =
   1.0000
   1.0000
   1.0000
   1.0000
   1.0000
   1.0000
```

At this level of accuracy, the results confirm that our algorithm is working.

However if the solution is displayed again using long format we get

```
» x'
ans =
    0.999999999999072
    1.0000000000267
    0.999999999818265
    1.00000000047487
    0.999999999473962
    1.00000000020787
```

which suggests that there is considerable build-up of roundoff error in this solution.

This loss of accuracy is due to the fact that the Hilbert matrix is a well known example of an *ill-conditioned* matrix. We shall discuss this property in a little more detail later. For now, it suffices to say that an ill-conditioned system is one where the accumulation of roundoff error can be severe. Another characterization is that the solution is highly sensitive to small changes in the data. (Of course, the effect of roundoff error is equivalent to having made small changes in the data.) In the case of Hilbert matrices, the severity of the ill-conditioning increases rapidly with the dimension of the matrix.

For the corresponding 10×10 Hilbert matrix system to that used in Example 5, the 'solution vector' obtained using the m-file above was

$$[1.0000, 1.0000, 1.0000, 1.0000, 0.9999, 1.0002, 0.9997, 1.0003, 0.9998, 1.0000]$$

showing very significant errors – especially bearing in mind that MATLAB is working internally with about 15 significant (decimal) figures.

Why do we believe these errors are caused by this notion of ill-conditioning rather than something inherent in the program?

MATLAB's rand function can be used to generate random matrices. Using the same approach as above we can choose the right-hand side vector to force the exact solution to be a vector of ones. The computed solution can then be compared with the exact one to get information about the performance of the algorithm. The commands listed below were used to repeat this experiment with 100 different matrices of each dimension from 4×4 up to 10×10:

```
» for n=4:10;
    for k=1:100
      A=−5+10*rand(n);
      b=sum(A')';
      x=gepp(A,b);
      E(k,n−3)=max(abs(x−1));
    end
  end
```

Each trial generates a random matrix with entries in the interval $[-5, 5]$. The matrix E contains the largest individual component error for each trial. Its largest entry is therefore the worst single component error in the complete set of 700 trials:

```
» max(max(E))
ans =
    3.0903e–012
```

This output shows that none of these randomly generated test examples generated errors that were especially large. Of course, some of these random matrices could be somewhat ill-conditioned themselves. The command

```
» sum(sum(E >1e-14))
```

counts the number of trials for which the worst error was greater than 10^{-14}. There were just 30 such out of the 700.

Note that MATLAB's sum function sums the columns of a matrix. Therefore the first sum in the command above generates a count of the number of cases where $E > 10^{-14}$ for each dimension, the second sum adds these to obtain the overall count.

Gauss elimination with partial pivoting appears to be a highly successful technique for solving linear systems.

7.2.2 Tridiagonal systems

In this section, we consider the important special case of tridiagonal systems of linear equations. We encountered these in both cubic spline interpolation and finite difference methods for boundary-value problems.

Recall that a tridiagonal matrix has its only nonzero entries on its main diagonal and immediately above and below this diagonal. We can economize on the storage of such a matrix by storing three vectors representing these diagonals of the complete matrix. We shall describe our generic tridiagonal system as

$$\begin{bmatrix} a_1 & b_1 & & & & \\ c_1 & a_2 & b_2 & & & \\ & c_2 & a_3 & b_3 & & \\ & & \ddots & \ddots & \ddots & \\ & & & c_{n-2} & a_{n-1} & b_{n-1} \\ & & & & c_{n-1} & a_n \end{bmatrix} \begin{bmatrix} x_1 \\ x_2 \\ x_3 \\ \vdots \\ x_{n-1} \\ x_n \end{bmatrix} = \begin{bmatrix} r_1 \\ r_2 \\ r_3 \\ \vdots \\ r_{n-1} \\ r_n \end{bmatrix} \tag{7.10}$$

where all the other entries in the matrix are zero.

At each step of the elimination, there is only one entry below the diagonal and so there is only one multiplier to compute. When a (multiple of a) row is subtracted from the one below it, the only positions affected are the diagonal and the immediate subdiagonal. The latter is of course going to be zeroed.

The effect of these observations is that the inner loops of the Gauss elimination algorithm, Algorithm 1, collapse to single operations.

Algorithm 4   Gauss elimination for a tridiagonal system

Input $n \times n$ tridiagonal matrix A as in (7.10), right-hand side n-vector **r**

Elimination

    for i=1:n−1

$$m := \frac{c_i}{a_i}; \ c_i := 0$$

$$a_{i+1} := a_{i+1} - mb_i$$

$$r_{i+1} := r_{i+1} - mr_i$$

    end

Back substitution

$$x_n = r_n/a_n$$

    for i=n−1:−1:1

$$x_i = (r_i - b_i x_{i+1})/a_i$$

    end

Output   Solution vector **x**

This algorithm is easy to program. By storing the tridiagonal matrix as a set of three vectors, the array-size limitations of the *Student Edition* of MATLAB can be overcome for such systems of equations. In Version 5 of the *Student Edition*, the largest square matrix permitted is 128×128 – which would not be enough for many boundary-value problems, for example. However, if the tridiagonal matrix is stored as 3 vectors this size limit rises to 16384×16384. (Even if the 3 vectors are stored as columns of another matrix a tridiagonal matrix of dimension 5460×5460 could be accommodated. This may be more convenient for passing such matrices into m-files.)

Example 6   For cubic spline interpolation with equally spaced knots, the tridiagonal system which must be solved to obtain the coefficients of the spline components has a matrix of the form

$$\begin{bmatrix} 4 & 1 & & & & \\ 1 & 4 & 1 & & & \\ & 1 & 4 & 1 & & \\ & & \ddots & \ddots & \ddots & \\ & & & 1 & 4 & 1 \\ & & & & 1 & 4 \end{bmatrix}$$

Solve the 5×5 system

$$\begin{bmatrix} 4 & 1 & & & \\ 1 & 4 & 1 & & \\ & 1 & 4 & 1 & \\ & & 1 & 4 & 1 \\ & & & 1 & 4 \end{bmatrix} \begin{bmatrix} x_1 \\ x_2 \\ x_3 \\ x_4 \\ x_5 \end{bmatrix} = \begin{bmatrix} 4.1 \\ 2.4 \\ 4.2 \\ 2.4 \\ 4.1 \end{bmatrix}$$

The elimination phase results in the bidiagonal (augmented) matrix

$$
\begin{bmatrix}
4 & 1 & & & & : & 4.1 \\
0 & 3.75 & 1 & & & : & 1.375 \\
 & 0 & 3.733333 & 1 & & : & 3.833333 \\
 & & 0 & 3.732143 & 1 & : & 1.373214 \\
 & & & 0 & 3.732057 & : & 3.732058
\end{bmatrix}
$$

from which we obtain the solution

$$x_5 = 3.732058/3.732057 = 1.000000$$
$$x_4 = (1.373214 - 1.000000)/3.732143 = 0.100000$$
$$x_3 = (3.833333 - 0.100000)/3.733333 = 1.000000$$
$$x_2 = (1.375 - 1.000000)/3.75 = 0.100000$$
$$x_1 = (4.1 - 0.100000)/4 = 1.000000$$

The important difference between this algorithm and Algorithm 1 is that it requires a much smaller amount of computation. For a square system, we stated earlier that approximately $n^3/3$ multiplications and $n^2/2$ divisions are needed. For a tridiagonal system, using Algorithm 4 reduces these counts to just $3n$ and $2n$ respectively.

Exercises: Section 7.2

1   Solve the system

$$
\begin{bmatrix}
1 & 2 & 3 & 4 \\
2 & 2 & 3 & 4 \\
3 & 3 & 3 & 4 \\
4 & 4 & 4 & 4
\end{bmatrix}
\begin{bmatrix}
w \\
x \\
y \\
z
\end{bmatrix}
=
\begin{bmatrix}
10 \\
11 \\
11 \\
12
\end{bmatrix}
$$

using Gauss elimination without pivoting.

2   Repeat Exercise 1, using partial pivoting.

3   Write a MATLAB program to solve a system of linear equations by Gauss elimination without pivoting. Test your program by re-solving Exercise 1.

4   Solve the Hilbert matrix systems (with exact solution vectors all ones) for dimensions $4, 5, \ldots, 10$ using Gauss elimination.

5   Repeat Exercise 4, using partial pivoting.

6   Write a program to solve a tridiagonal system using Gauss elimination. Test it on the 50×50 system with diagonal entries $1, 2, \ldots, 50$, subdiagonal and superdiagonal entries all 1, and with exact solution vector $[1, 1, 1, \ldots, 1]'$.

7   Write a program to solve a linear 2-point boundary-value problem using the second-order finite difference method of Section 6.6. Use your tridiagonal-solver to solve the linear system of equations that results. Use it to solve the Bessel equation $x^2 y'' + x y' + (x^2 - 1)y = 0$ with $y(1) = 1$, $y(20) = 0$ using 190 steps. (Note this cannot be solved using the *Student Edition* of MATLAB as a full matrix. The use of a tridiagonal-solver is critical to this solution.)

7.3 *LU* factorization: iterative refinement

One of the difficulties with the Gauss elimination approach to the solution of linear equations lies in the accumulation of roundoff errors. We saw in Section 7.2 that the use of pivoting can improve matters but, even then, if the system is ill-conditioned, large relative errors can result.

It can be helpful to look at the vector of *residuals* for a computed solution $\tilde{\mathbf{x}}$ to a system

$$A\mathbf{x} = \mathbf{b} \tag{7.11}$$

We define the residual vector by

$$\mathbf{r} = \mathbf{b} - A\tilde{\mathbf{x}} \tag{7.12}$$

The components of $\mathbf{r}$ given by

$$r_k = b_k - (a_{k1}\tilde{x}_1 + a_{k2}\tilde{x}_2 + \cdots + a_{kn}\tilde{x}_n)$$

for $k = 1, 2, \ldots, n$ provide a measure (although not always a very good one) of the extent to which we have failed to satisfy the equations. Now, if we can also solve the system

$$A\mathbf{y} = \mathbf{r} \tag{7.13}$$

then, by adding this solution $\mathbf{y}$ to the original computed solution $\tilde{\mathbf{x}}$, we obtain

$$A(\tilde{\mathbf{x}} + \mathbf{y}) = A\tilde{\mathbf{x}} + A\mathbf{y} - A\tilde{\mathbf{x}} + r$$
$$= A\tilde{\mathbf{x}} + b - A\tilde{\mathbf{x}} = b$$

Thus if we can compute $\mathbf{r}$ and solve (7.13) *exactly*, then $\tilde{\mathbf{x}} + \mathbf{y}$ is the exact solution of the original system (7.11). Of course, in practice, we cannot solve this system exactly but we might anticipate that $\tilde{\mathbf{x}} + \tilde{\mathbf{y}}$ will be an improved solution. (Here $\tilde{\mathbf{y}}$ represents the computed solution of (7.13).) This suggests a possible iterative method, known as *iterative refinement*, for solving a system of linear equations.

It has already been mentioned that Gauss elimination can be used to solve several systems with the same coefficient matrix but different right-hand sides simultaneously. Essentially the only change is that the operations performed on the elements of the right-hand side *vector* must now be applied to the *rows* of the right-hand side *matrix*. Unfortunately, however, we do not know the residual vector at the time of the original solution. That is to say that (7.13) must be solved *after* the solution $\tilde{\mathbf{x}}$ to (7.11) has been obtained. We wish to avoid repeating all the work of the original solution in order to solve the second system. For this reason it is desirable to keep track of the multipliers used in the Gauss elimination process. This leads us to what is called *LU factorization* of the matrix A in which we obtain two matrices L, a lower triangular matrix, and U, an upper triangular matrix, with the property that

$$A = LU \tag{7.14}$$

This process requires no more than that we store the multipliers used in Gauss elimination. To illustrate this, we consider again the first step of the Gauss elimination phase where we generate the matrix

$$A^{(1)} = \begin{bmatrix} a_{11} & a_{12} & a_{13} & \cdots & a_{1n} \\ 0 & a'_{22} & a'_{23} & \cdots & a'_{2n} \\ 0 & a'_{32} & a'_{33} & \cdots & a'_{3n} \\ \vdots & & & & \\ 0 & a'_{n2} & a'_{n3} & \cdots & a'_{nn} \end{bmatrix}$$

using the operations

$$a'_{jk} = a_{jk} - \frac{a_{j1}}{a_{11}} a_{1k} = a_{jk} - m_{j1} a_{1k}$$

(See (7.6) and (7.7).) Now define

$$M_1 = \begin{bmatrix} 1 & 0 & \cdots & \cdots & 0 \\ m_{21} & 1 & 0 & \cdots & 0 \\ m_{31} & 0 & 1 & & \\ \vdots & \vdots & & \ddots & \\ m_{n1} & 0 & & & 1 \end{bmatrix} \tag{7.15}$$

The effect of premultiplying any matrix by M_1 is to add m_{j1} times the first row to the j-th row. Therefore, multiplying $A^{(1)}$ by M_1 would exactly reverse the elimination that has been performed, in other words

$$M_1 A^{(1)} = A \tag{7.16}$$

A similar equation holds for each step of the elimination phase. It follows that denoting the final upper triangular matrix produced by Algorithm 1 by U and defining the lower triangular matrix

$$L = \begin{bmatrix} 1 & & & & & \\ m_{21} & 1 & & & & \\ m_{31} & m_{32} & 1 & & & \\ m_{41} & m_{42} & m_{43} & \ddots & & \\ \vdots & \vdots & \vdots & & 1 & \\ m_{n1} & m_{n2} & m_{n3} & \cdots & m_{n,n-1} & 1 \end{bmatrix} \tag{7.17}$$

where the m_{ji} are just the multipliers used in Algorithm 1, we get

$$LU = A$$

as desired.

As an algorithm, this corresponds to simply overwriting the subdiagonal parts of A with the multipliers so that both the upper and lower triangles can be stored in the same local (matrix) variable. We summarize this as follows.

Algorithm 5   LU factorization of a square matrix

Input $n \times n$ matrix A
Factorization
  for i=1:n-1
    for j=i+1:n

$$a_{ji} := \frac{a_{ji}}{a_{ii}};$$

      for k=i+1:n

$$a_{jk} := a_{jk} - a_{ji}a_{ik}$$

    end
   end
  end

Output   Modified matrix A containing upper and lower triangular factors
  L is the lower triangle of A with unit diagonal elements
  U is the upper triangle of A

It should be observed that this is not the only LU factorization of A. The two factors can be scaled appropriately to allow (almost) any choice for the diagonal of one of the factors. This particular choice, which is the most commonly used for general matrices, is called the *Dolittle reduction* of A. An alternative is to scale the rows of U and columns of L so that U has ones on its diagonal. This is called the *Crout reduction*. In the special case of a symmetric matrix A, choosing $U = L'$ so that $A = LL'$ has some advantages. This is the *Cholesky factorization*. We shall deal only with the Dolittle reduction.

This algorithm is essentially the same as Algorithm 1, except that the multipliers are stored in the lower triangle and the operations on the right-hand side of our linear system have been omitted. So how do we use Algorithm 5 to solve the original system?

The process is easily summarized as consisting of three steps

1 Factor the matrix $A = LU$ as in Algorithm5
2 Solve the lower triangular system $Lz = \mathbf{b}$ using *forward substitution*
3 Solve the upper triangular system $U\mathbf{x} = \mathbf{z}$ using back substitution as in Algorithm 2

Note that the forward substitution algorithm is the natural equivalent of back substitution. It is slightly simplified by the fact that the diagonal entries of L are all one so that the divisions are unnecessary.

First we establish that this process really solves the system. We have

$$A\mathbf{x} = LU\mathbf{x} = L\mathbf{z} = \mathbf{b}$$

as required.

The second question that arises naturally is: how does this 3-stage process compare with Gauss elimination in terms of efficiency? The answer is simple: the total numbers of floating-point operations are identical. All that has changed is the *order* in which some of them are performed.

So what is the advantage? Consider the iterative refinement process that we discussed above. The initial solution $\tilde{\mathbf{x}}$ is computed using $O(n^3)$ floating-point operations. Then the residual vector is computed as

$$\mathbf{r} = \mathbf{b} - A\tilde{\mathbf{x}}$$

after which the correction can be computed by solving

$$Lz = \mathbf{r}$$
$$Uy = \mathbf{z}$$

and setting

$$\mathbf{x} = \tilde{\mathbf{x}} + \mathbf{y}$$

The important difference is that the factorization does not need to be repeated. The computation of the iterative refinement correction consists only of forward and back substitution which are each $O(n^2)$ floating-point operations. For large systems this represents a significant saving.

Example 7   Solve the system

$$\begin{bmatrix} 7 & -7 & 1 \\ -4 & 4 & -1 \\ 7 & 7 & -4 \end{bmatrix} \begin{bmatrix} x_1 \\ x_2 \\ x_3 \end{bmatrix} = \begin{bmatrix} 1 \\ -1 \\ 10 \end{bmatrix}$$

using LU factorization and iterative refinement

First we note that this is the same system that we used in Example 3. From the multipliers used there, we obtain the factorization

$$\begin{bmatrix} 7 & -7 & 1 \\ -4 & 4 & -1 \\ 7 & 7 & -3 \end{bmatrix} = \begin{bmatrix} 1 & 0 & 0 \\ -0.5714 & 1 & 0 \\ 1 & 70000 & 1 \end{bmatrix} \begin{bmatrix} 7 & -7 & 1 \\ 0 & 0.0002 & -0.4286 \\ 0 & 0 & 29997 \end{bmatrix}$$

Solving $Lz = [1, -1, 10]'$ we obtain $\mathbf{z} = [1, -0.4286, 30011]'$ which is of course the right-hand side at the end of the elimination phase in Example 3. Now solving $U\mathbf{x} = \mathbf{z}$, we get our computed solution

$$\tilde{\mathbf{x}} = \begin{bmatrix} 1.0714 \\ 1.0715 \\ 1.0005 \end{bmatrix}$$

as before.

The residuals are now given by

$$\mathbf{r} = \begin{bmatrix} 1 \\ -1 \\ 10 \end{bmatrix} - \begin{bmatrix} 7 & -7 & 1 \\ -4 & 4 & -1 \\ 7 & 7 & -4 \end{bmatrix} \begin{bmatrix} 1.0714 \\ 1.0715 \\ 1.0005 \end{bmatrix} = \begin{bmatrix} 0.0002 \\ 0.0001 \\ -0.9983 \end{bmatrix}$$

For the iterative refinement, we first solve

$$\begin{bmatrix} 1 & 0 & 0 \\ -0.5714 & 1 & 0 \\ 1 & 70000 & 1 \end{bmatrix} \mathbf{z} = \begin{bmatrix} 0.0002 \\ 0.0001 \\ -0.9983 \end{bmatrix}$$

to obtain $z_1 = 0.0002$, $z_2 = 0.0001 - (-0.5714)(0.0002) = 0.0002$ and $z_3 = -0.9983 - (70000)(0.0002) - 0.0002 = -14.9985$. Then solving $U\mathbf{y} = [0.0002, 0.0002, -14.9985]$ yields

$$y_3 = -14.9985/29997 = -0.0005$$
$$y_2 = (0.0002 - (-0.4286)(-0.0005))/0.0002 = -0.0715$$
$$y_1 = (0.0002 - (-7)(-0.0715) - (-0.0005))/7 = -0.0714$$

Finally

$$\tilde{\mathbf{x}} + y = \begin{bmatrix} 1.0714 \\ 1.0715 \\ 1.0005 \end{bmatrix} + \begin{bmatrix} -0.0714 \\ -0.0715 \\ -0.0005 \end{bmatrix} = \begin{bmatrix} 1.0000 \\ 1.0000 \\ 1.0000 \end{bmatrix}$$

Full working accuracy has been restored in the solution by the iterative refinement.

Although no pivoting was used in Example 7, it is not too difficult to incorporate partial pivoting into the *LU* factorization algorithm. It is necessary to keep track of the order in which the rows were used as pivots in order to adjust the right-hand side appropriately. We shall not discuss the details of this implementation here. As a general rule, iterative refinement is usually of little value if *LU* factorization with partial pivoting is used, *unless* the residuals are computed to greater precision than that which is used for the original solution. In an environment such as MATLAB where all floating-point arithmetic uses IEEE double-precision floating-point it is unlikely to offer any significant advantage.

When pivoting is used, the factorization becomes equivalent to finding lower and upper triangular factors, and a *permutation matrix* P such that

$$LU = PA$$

in which case the forward and back substitution must take this into account by solving not $A\mathbf{x} = \mathbf{b}$ but $PA\mathbf{x} = P\mathbf{b}$.

LU factorization with partial pivoting is implemented in MATLAB as the function lu. It can return either 2 or 3 matrices. With just two outputs, it returns the upper triangular factor and a 'lower triangular' factor with its rows interchanged. If three outputs are asked for then the lower triangular factor is genuinely a lower triangle and the third argument is the permutation matrix.

For example with the same matrix as we used in Example 7, we can use

[L,U]=lu(A)

to obtain the output

L =

1.0000	0	0
−0.5714	0	1.0000
1.0000	1.0000	0

U =

7.0000	−7.0000	1.0000
0	14.0000	−5.0000
0	0	−0.4286

where we see that the second and third rows of *L* have been interchanged to reflect the pivoting strategy. Note that

$$\begin{bmatrix} 1.0000 & 0 & 0 \\ -0.5714 & 0 & 1.0000 \\ 1.0000 & 1.0000 & 0 \end{bmatrix} \begin{bmatrix} 7.0000 & -7.0000 & 1.0000 \\ 0 & 14.0000 & -5.0000 \\ 0 & 0 & -0.4286 \end{bmatrix}$$

$$= \begin{bmatrix} 7.0 & -7.0 & 1.0 \\ -3.999\,8 & 3.999\,8 & -1.0 \\ 7.0 & 7.0 & -4.0 \end{bmatrix}$$

which is the original matrix *A* with small roundoff errors. Of course the MATLAB product *LU* would be much more accurate since the elements of the factors would be stored to greater accuracy as is demonstrated by

» L*U

ans =

```
 7 −7  1
−4  4 −1
 7  7 −4
```

Alternatively, the command

[L,U,P]=lu(A)

yields

L =

1.0000	0	0
1.0000	1.0000	0
−0.5714	0	1.0000

```
U =
   7.0000   -7.0000    1.0000
        0   14.0000   -5.0000
        0         0   -0.4286
P =
   1   0   0
   0   0   1
   0   1   0
```

The permutation matrix P carries the information on the order in which the rows were used. This time, we see that the product of L and U is

```
» L*U
ans =
   7 -7  1
   7  7 -4
  -4  4 -1
```

which is the original matrix A with its second and third rows interchanged. This interchange is precisely the effect of multiplying by the permutation matrix P:

```
» P*A
ans =
   7 -7  1
   7  7 -4
  -4  4 -1
```

To complete the solution of the system in Example 7 we can then find $\mathbf{z}$ such that $L\mathbf{z} = P\mathbf{b}$ and $\mathbf{x}$ such that $U\mathbf{x} = \mathbf{z}$. These operations yield the correct answers to full machine accuracy in this case.

The advantage of the LU factorization is that if we now wished to solve another system with the same coefficient matrix, the elimination (factorization) phase does not need to be repeated. For example if we wish to solve

$$\begin{bmatrix} 7 & -7 & 1 \\ -4 & 4 & -1 \\ 7 & 7 & -4 \end{bmatrix} \begin{bmatrix} x_1 \\ x_2 \\ x_3 \end{bmatrix} = \begin{bmatrix} 15 \\ -9 \\ -4 \end{bmatrix}$$

then we just multiply the right-hand side by the permutation matrix to get $[15, -4, -9]'$ and then use forward and back substitution to obtain

$$\mathbf{z} = \begin{bmatrix} 15 \\ -19 \\ -0.4286 \end{bmatrix} \text{ and } \mathbf{x} = \begin{bmatrix} 1 \\ -1 \\ 1 \end{bmatrix}$$

LU factorization is the basic method used by MATLAB's \ linear equation solution operator for square systems that are not ill-conditioned.

**Exercises:
Section 7.3**

1 Obtain LU factors of

$$A = \begin{bmatrix} 7 & 8 & 8 \\ 6 & 5 & 4 \\ 1 & 2 & 3 \end{bmatrix}$$

using 4 decimal place arithmetic. Use these factors to solve the following system of equations

$$A \begin{bmatrix} x \\ y \\ z \end{bmatrix} = \begin{bmatrix} 7 \\ 5 \\ 2 \end{bmatrix}$$

with iterative refinement.

2 Write a program to perform forward substitution. Test it by solving the following system

$$\begin{bmatrix} 1 & & & & \\ 1 & 1 & & & \\ 2 & 3 & 1 & & \\ 4 & 5 & 6 & 1 & \\ 7 & 8 & 9 & 0 & 1 \end{bmatrix} \begin{bmatrix} x_1 \\ x_2 \\ x_3 \\ x_4 \\ x_5 \end{bmatrix} = \begin{bmatrix} 0.1 \\ 0.4 \\ 1.6 \\ 5.6 \\ 8.5 \end{bmatrix}$$

3 Write a MATLAB m-file to solve a square system of equations using the built-in LU factorization function and your forward and back substitution programs. Use these to solve the Hilbert systems of dimensions 6 through 10 with solution vectors all ones.

4 For the solutions of Exercise 3, test whether iterative refinement offers any improvement.

5 Solve the 10×10 Vandermonde system $V\mathbf{x} = \mathbf{b}$ where $v_{ij} = i^{j-1}$ and $b_i = (i/4)^{10}$, using LU factorization.

7.4 Iterative methods

In this section, we again consider the solution of a basic linear system

$$A\mathbf{x} = \mathbf{b}$$

and consider alternatives to the direct methods of Gauss elimination and LU factorization. Especially if the dimension of the system is large and if the system is *sparse* (which is to say that many of the coefficients are zero) then iterative methods can be an attractive option.

In this section, we consider just two of these iterative techniques. These two form the basis of a family of methods which are designed either to accelerate the convergence or to suit some particular computer architecture. Both our methods are based on rearranging the original system of equations in the form

$$
\begin{aligned}
x_1 &= [b_1 - (a_{12}x_2 + a_{13}x_3 + \cdots + a_{1n}x_n)]/a_{11} \\
x_2 &= [b_2 - (a_{21}x_1 + a_{23}x_3 + \cdots + a_{2n}x_n)]/a_{22} \\
&\cdots \quad \cdots \quad \cdots \quad \cdots \quad \cdots \quad \cdots \\
x_n &= \left[b_n - \left(a_{n1}x_1 + a_{n2}x_2 + \cdots + a_{n,n-1}x_{n-1}\right)\right]/a_{nn}
\end{aligned}
\tag{7.18}
$$

which is the result of solving the i-th equation for x_i in terms of the remaining unknowns. There is an implicit assumption here that all diagonal elements of A are nonzero. (It is always possible to reorder the equations of a nonsingular system to ensure this condition is satisfied.)

In matrix notation, (7.18) is equivalent to writing

$$
\mathbf{x} = D^{-1}[\mathbf{b} - (L + U)\mathbf{x}]
\tag{7.19}
$$

where

$$
A = L + D + U
\tag{7.20}
$$

and L is strictly lower triangular, D is diagonal, and U is strictly upper triangular:

$$
l_{ij} = \begin{cases} a_{ij} & \text{if } i > j \\ 0 & \text{otherwise} \end{cases}
$$

$$
d_{ii} = a_{ii}, \; d_{ij} = 0 \text{ otherwise}
$$

$$
u_{ij} = \begin{cases} a_{ij} & \text{if } i < j \\ 0 & \text{otherwise} \end{cases}
$$

This rearrangement of the original system lends itself to an iterative treatment.

For the *Jacobi iteration*, we generate the next estimated solution vector from the current one by substituting the current component values in the right-hand side of (7.18) to obtain the next iterates. In matrix terms, we set

$$
\mathbf{x}^{(k+1)} = D^{-1}\left[\mathbf{b} - (L + U)\mathbf{x}^{(k)}\right]
\tag{7.21}
$$

where the superscript represents the iteration number so that $\mathbf{x}^{(k)}$ is the k-th (vector) iterate. In component terms, we have

$$
\begin{aligned}
x_1^{(k+1)} &= \left[b_1 - \left(a_{12}x_2^{(k)} + a_{13}x_3^{(k)} + \cdots + a_{1n}x_n^{(k)}\right)\right]/a_{11} \\
x_2^{(k+1)} &= \left[b_2 - \left(a_{21}x_1^{(k)} + a_{23}x_3^{(k)} + \cdots + a_{2n}x_n^{(k)}\right)\right]/a_{22} \\
&\cdots \quad \cdots \quad \cdots \quad \cdots \quad \cdots \\
x_n^{(k+1)} &= \left[b_n - \left(a_{n1}x_1^{(k)} + a_{n2}x_2^{(k)} + \cdots + a_{n,n-1}x_{n-1}^{(k)}\right)\right]/a_{nn}
\end{aligned}
\tag{7.22}
$$

Example 8   **Perform the first 3 Jacobi iterations for the solution of the system**

$$48c_1 + 13c_2 = -0.0420$$
$$13c_1 + 56c_2 + 15c_3 = -0.0306$$
$$15c_2 + 64c_3 = -0.0237$$

(This system was derived in Chapter 4 (p. 107), for fitting a cubic spline to the square-root function.)

Rearranging the equations as in (7.18) and (7.19), we have

$$c_1 = \frac{-0.0420 - 13c_2}{48}$$
$$c_2 = \frac{-0.0306 - 13c_1 - 15c_3}{56}$$
$$c_3 = \frac{-0.0237 - 15c_2}{64}$$

and taking the initial guess $\mathbf{c} = \mathbf{0}$, we get the next iterates

$$c_1 = \frac{-0.0420}{48} = -0.000875$$
$$c_2 = \frac{-0.0306}{56} = -0.000546$$
$$c_3 = \frac{-0.0237}{64} = -0.000370$$

The next two iterations yield

$$c_1 = -0.000727, \quad c_2 = -0.000244, \quad c_3 = -0.000362$$

and

$$c_1 = -0.000809, \quad c_2 = -0.000281, \quad c_3 = -0.000313$$

which are steadily, if slowly, approaching the true solution $[-0.000799, -0.000279, -0.000305]$.

Clearly for this small system, we could obtain the accurate solution with less effort than has already been expended. The purpose of the example is simply to illustrate the process.

An obvious question arises from this Jacobi iteration. Once $x_1^{(k+1)}$ has been obtained in the first of equations (7.22), why not use this supposedly better estimate in place of $x_1^{(k)}$ in the remainder of the updates in (7.22)? Similar observations apply to all the updates: why not use the most recent information at all times? The answer is that we can – the result is the *Gauss–Seidel iteration*:

$$x_1^{(k+1)} = \left[b_1 - \left(a_{12}x_2^{(k)} + a_{13}x_3^{(k)} + \cdots + a_{1n}x_n^{(k)}\right)\right]/a_{11}$$

$$x_2^{(k+1)} = \left[b_2 - \left(a_{21}x_1^{(k+1)} + a_{23}x_3^{(k)} + \cdots + a_{2n}x_n^{(k)}\right)\right]/a_{22}$$

$$\cdots \qquad \cdots \qquad \cdots \qquad \cdots \qquad \cdots \qquad \cdots$$

$$x_n^{(k+1)} = \left[b_n - \left(a_{n1}x_1^{(k+1)} + a_{n2}x_2^{(k+1)} + \cdots + a_{n,n-1}x_{n-1}^{(k+1)}\right)\right]/a_{nn}$$

(7.23)

or, in matrix terms,

$$\mathbf{x}^{(k+1)} = D^{-1}\left[\mathbf{b} - \left(L\mathbf{x}^{(k+1)} + U\mathbf{x}^{(k)}\right)\right]$$

(7.24)

Example 9   Repeat Example 8 using the Gauss–Seidel iteration

With the same initial guess, the first iteration is now

$$c_1 = \frac{-0.0420}{48} = -0.000875$$

$$c_2 = \frac{-0.0306 - 13(-0.000875)}{56} = -0.000343$$

$$c_3 = \frac{-0.0237 - 15(-0.000343)}{64} = -0.000290$$

The next two iterations then produce the estimates

$$c_1 = -0.000782, \quad c_2 = -0.000259, \quad c_3 = -0.000310$$

$$c_1 = -0.000805, \quad c_2 = -0.000277, \quad c_3 = -0.000305$$

which are much closer to the true solution than the Jacobi iterates for the same computational effort.

Either of the iterations (7.22) or (7.23) uses approximately n^2 multiplications and n divisions per iteration. Comparing this with the operation counts for Gauss elimination, we see that these iterative methods are likely to be computationally less expensive if they will converge in fewer than about $n/3$ iterations. Obviously for small systems this is unlikely, but for large sparse systems this may often be the case.

This raises the issue of when Jacobi or Gauss–Seidel iterations will converge. Examples 8 and 9 seem to exhibit convergence but is this typical? Certainly it is not universal, as the following slight modification to Example 9 shows. If we reorder the equations, by interchanging the first two so that the system is

$$13c_1 + 56c_2 + 15c_3 = -0.0306$$
$$48c_1 + 13c_2 = -0.0420$$
$$15c_2 + 64c_3 = -0.0237$$

then the first two Gauss–Seidel iterations give us

$$c_1 = -0.00235, \quad c_2 = -0.00545, \quad c_3 = -0.00165$$
$$c_1 = -0.0239, \quad c_2 = -0.0851, \quad c_3 = -0.0203$$

which are clearly moving rapidly *away* from the solution.

What is the important difference between these two systems?

In the original ordering of the equations, the matrix of coefficients is *diagonally dominant* which is to say that each diagonal element is greater than the (absolute) sum of the other entries in its row, or

$$|a_{ii}| > \sum_{j \neq i} |a_{ij}|$$

for each *i*. For our example, we have $48 > 13, 56 > 13 + 15$, and $64 > 15$ whereas in the reordered system $13 < 56 + 15$ and $13 < 48$.

The simplest conditions under which both Jacobi and Gauss–Seidel iterations can be proved to converge is when the coefficient matrix is diagonally dominant. The details of these proofs are beyond our present scope. They are essentially multivariable versions of the conditions for convergence of fixed-point (function) iteration which were discussed for a single equation in Chapter 2. The diagonal dominance of the matrix ensures that the various (multivariable) derivatives of the iteration functions (7.21) and (7.24) are all less than unity.

The apparently more rapid convergence of the Gauss–Seidel iteration in Example 9 is not universal but it is common. However, we should observe that on a parallel computer, the Gauss–Seidel iteration would not be as useful. The updates of a Jacobi iteration can all be performed simultaneously, whereas for Gauss–Seidel, $x_2^{(k+1)}$ cannot be computed until after $x_1^{(k+1)}$ is known. There are other iterative schemes which can be employed to try to take some advantage of the ideas of the Gauss–Seidel iteration in a parallel computing environment.

This point is readily borne out by the efficient implementation of the Jacobi iteration in MATLAB where we can take advantage of the matrix operations.

Program   **MATLAB code for fixed number of Jacobi iterations**

```
function s=Jacobit(A,b,Nits)
% Performs Nits Jacobi iterations for the solution of
% Ax=b
% A is assumed square
```

```
D=diag(A);
n=size(A,1);
A_D=A-diag(D); % This is L+U
x=zeros(n,1);
for k=1:Nits
    x=(b-A_D*x)./D;
    s(:,k)=x;
end
```

Note the ease of implementation that MATLAB's matrix arithmetic allows for the Jacobi iteration. The same simplicity could not be achieved for Gauss–Seidel because of the need to compute each component in turn. The implicit inner loop would need to be 'unrolled' into its component form.

Example 10 Perform the first six Jacobi iterations for a randomly generated diagonally dominant 10×10 system of linear equations

One way to generate a suitable system is

» A=10*eye(10);
» A=A+rand(10);
» b=sum(A')';

Since the random numbers are all in $[0, 1]$, the construction of A guarantees its diagonal dominance. The right-hand side is chosen to make the solution a vector of ones. Using the program listed above we get the table of results below.

» S=jacobit(A,b,6);

Iteration	1	2	3	4	5	6
x_1	1.3539	0.8612	1.0583	0.9761	1.0099	0.9959
x_2	1.3901	0.8320	1.0684	0.9717	1.0116	0.9952
x_3	1.5581	0.7752	1.0931	0.9617	1.0158	0.9935
x_4	1.4068	0.8362	1.0679	0.9721	1.0115	0.9953
x_5	1.3216	0.8668	1.0541	0.9776	1.0092	0.9962
x_6	1.4951	0.8024	1.0817	0.9663	1.0139	0.9943
x_7	1.4474	0.8210	1.0742	0.9695	1.0126	0.9948
x_8	1.4179	0.8173	1.0749	0.9692	1.0127	0.9948
x_9	1.3646	0.8458	1.0631	0.9739	1.0107	0.9956
x_{10}	1.4446	0.8168	1.0757	0.9688	1.0128	0.9947

The iterates are gradually approaching the imposed solution

The Jacobi iteration is particularly easy to program, but the Gauss–Seidel is typically more efficient. The results of using five Gauss–Seidel iterations for the system generated in Example 10 are tabulated below.

Iteration	1	2	3	4	5
x_1	1.3539	0.9712	1.0005	1.0002	1.0000
x_2	1.3611	0.9625	0.9984	1.0003	1.0000
x_3	1.3660	0.9908	0.9980	1.0001	1.0000
x_4	1.2535	0.9959	0.9981	1.0000	1.0000
x_5	1.1166	1.0030	0.9988	1.0000	1.0000
x_6	1.2196	1.0150	0.9987	0.9999	1.0000
x_7	1.0145	1.0093	0.9999	0.9999	1.0000
x_8	0.9796	1.0075	1.0003	1.0000	1.0000
x_9	0.9321	1.0023	1.0002	1.0000	1.0000
x_{10}	0.9253	1.0027	1.0003	1.0000	1.0000

It is apparent that the Gauss–Seidel iteration has given greater accuracy much more quickly.

Exercises: Section 7.4

1 Perform the first three Jacobi iterations on the system

$$\begin{bmatrix} 4 & 1 & & & \\ 1 & 4 & 1 & & \\ & 1 & 4 & 1 & \\ & & 1 & 4 & 1 \\ & & & 1 & 4 \end{bmatrix} \begin{bmatrix} x_1 \\ x_2 \\ x_3 \\ x_4 \\ x_5 \end{bmatrix} = \begin{bmatrix} 4.1 \\ 2.4 \\ 4.2 \\ 2.4 \\ 4.1 \end{bmatrix}$$

used in Example 6 on tridiagonal systems.

2 Repeat Exercise 1 for the Gauss–Seidel iteration. Which appears better?

3 Write a program to perform a fixed number of Gauss–Seidel iterations for the solution of a system $A\mathbf{x} = \mathbf{b}$. Test it on the system of Exercise 1.

4 Generate a random diagonally dominant matrix of dimension 20×20 and right-hand sides for true solutions (a) $x_i = 1$ for all i and (b) $x_i = (-1)^i$. Perform the first ten iterations of both Jacobi and Gauss–Seidel for each case. Compare your results.

5 Modify your code for the Gauss–Seidel iteration to continue until the maximum absolute difference between components of $\mathbf{x}^{(k)}$ and $\mathbf{x}^{(k+1)}$ is smaller than a given tolerance. Use your program to solve your example of Exercise 4 with accuracy 10^{-6}.

7.5 Linear least squares approximation

In this section, we return to one of the major themes of the book, the approximation of functions. We present only a brief introduction to this important topic which is the basis of many function approximation routines as well as linear regression in statistics. The solution to linear least squares approximation is an important application of the solution of systems of linear equations and leads to other interesting ideas of numerical linear algebra as we shall see shortly.

In Chapter 4, we pointed out some of the potential drawbacks of polynomial interpolation in the situation where the number of data points is large and the data are themselves subject to experimental or some other source of error. The idea behind least squares approximation is that we find an approximating function p, say, from some particular class, that minimizes the sum of the squares of the errors. For most of this section, we shall use polynomials of specified maximum degree as the approximating functions. There are very important examples of least squares approximations using functions other than polynomials. We shall discuss a few of these briefly later.

We recall from Section 1.4 that the continuous L_2 or *least squares*, metric is defined for the interval $[a, b]$ by

$$L_2(f, p) = ||f - p||_2 = \sqrt{\int_a^b |f(x) - p(x)|^2 dx} \qquad (7.25)$$

The *continuous* least squares approximation problem is to find the function p from the admissible set which minimizes this quantity. (In this case the 'sum' of the squares of the errors is a continuous sum, or integral.)

In a similar manner, if we have values of the function f at a set of points x_i for $i = 1, 2, \ldots, N$ then the corresponding *discrete* least squares metric is defined by

$$D_2(f, p) = \sqrt{\sum_{i=1}^N |f(x_i) - p(x_i)|^2} \qquad (7.26)$$

and the discrete least squares problem is to find that function p, again from some specified class, which minimizes this metric.

Initially, we restrict to polynomial approximations of degree not more than some M. Then we must find the coefficients of

$$p(x) = a_0 + a_1 x + a_2 x^2 + \cdots + a_M x^M \qquad (7.27)$$

which minimizes $L_2(f, p)$ or $D_2(f, p)$.

We begin with the continuous case and denote $(L_2(f, p))^2$ by $F(a_0, a_1, \ldots, a_M)$. Clearly the choice of $a_0, a_1, \ldots, a_M$ which minimizes F also minimizes $L_2(f, p)$. Now, by definition

$$F(a_0, a_1, \ldots, a_M) = \int_a^b \left[f(x) - \left(a_0 + a_1 x + a_2 x^2 + \cdots + a_M x^M \right) \right]^2 dx \qquad (7.28)$$

Setting each of the partial derivatives of F to zero, we obtain the following necessary conditions for the solution. These form a system of linear equations for the coefficients.

$$\frac{\partial F}{\partial a_0} = -2\int_a^b \left[f(x) - \left(a_0 + a_1 x + a_2 x^2 + \cdots + a_M x^M\right)\right]dx = 0$$

$$\frac{\partial F}{\partial a_1} = -2\int_a^b x\left[f(x) - \left(a_0 + a_1 x + a_2 x^2 + \cdots + a_M x^M\right)\right]dx = 0$$

$$\cdots \quad \cdots \quad \cdots \quad \cdots \quad \cdots \quad \cdots$$

$$\frac{\partial F}{\partial a_M} = -2\int_a^b x^M\left[f(x) - \left(a_0 + a_1 x + a_2 x^2 + \cdots + a_M x^M\right)\right]dx = 0$$

Dividing each of these by 2 and we can rewrite this system as

$$c_0 a_0 + c_1 a_1 + \cdots + c_M a_M = b_0$$
$$c_1 a_0 + c_2 a_1 + \cdots + c_{M+1} a_M = b_1$$
$$c_2 a_0 + c_3 a_1 + \cdots + c_{M+2} a_M = b_2 \qquad (7.29)$$
$$\cdots \quad \cdots \quad \cdots \quad = \cdots$$
$$c_M a_0 + c_{M+1} a_1 + \cdots + c_{2M} a_M = b_M$$

where the coefficients c_k and b_k for appropriate values of k are given by

$$c_k = \int_a^b x^k dx, \quad b_k = \int_a^b x^k f(x)dx \qquad (7.30)$$

In matrix terms the system is

$$\begin{bmatrix} c_0 & c_1 & c_2 & \cdots & c_M \\ c_1 & c_2 & \cdots & c_M & c_{M+1} \\ c_2 & \cdots & c_M & c_{M+1} & c_{M+2} \\ \cdots & & \cdots & & \cdots \\ c_M & c_{M+1} & c_{M+2} & \cdots & c_{2M} \end{bmatrix} \begin{bmatrix} a_0 \\ a_1 \\ a_2 \\ \vdots \\ a_M \end{bmatrix} = \begin{bmatrix} b_0 \\ b_1 \\ b_2 \\ \vdots \\ b_M \end{bmatrix} \qquad (7.31)$$

Note: The special structure of the matrix with constant entries on each 'reverse diagonal'. Such matrices are often called *Hankel matrices*.

In the discrete case a similar system would be obtained. The only difference being that the coefficients would be defined by the discrete analogues of (7.30):

$$c_k = \sum_{i=0}^N x_i^k, \quad b_k = \sum_{i=0}^N x_i^k f(x_i) \qquad (7.32)$$

In either case therefore the least squares approximation problem is reduced to a linear system of equations, called the *normal equations*. The matrix of coefficients is necessarily nonsingular (provided the data points are distinct in the discrete case) and so the problem has a unique solution. The reason this process is called *linear least squares* is that the approximating function is a linear combination of the basis functions, in this case the monomials $1, x, x^2, \ldots, x^M$.

Example 11 **Find the continuous least squares cubic approximation to** $\sin x$ **on** $[0, \pi]$

We must minimize

$$F(a_0, a_1, a_2, a_3) = \int_0^\pi \left(\sin x - a_0 - a_1 x - a_2 x^2 - a_3 x^3 \right) dx$$

The coefficients of the normal equations are given by

$$c_k = \int_0^\pi x^k dx = \frac{\pi^{k+1}}{k+1}$$

$$b_k = \int_0^\pi x^k \sin x \, dx$$

so that $b_0 = 2$, $b_1 = \pi$, $b_2 = \pi^2 - 4$, and $b_3 = \pi(\pi^2 - 6)$. The normal equations are therefore

$$\begin{bmatrix} \pi & \pi^2/2 & \pi^3/3 & \pi^4/4 \\ \pi^2/2 & \pi^3/3 & \pi^4/4 & \pi^5/5 \\ \pi^3/3 & \pi^4/4 & \pi^5/5 & \pi^6/6 \\ \pi^4/4 & \pi^5/5 & \pi^6/6 & \pi^7/7 \end{bmatrix} \begin{bmatrix} a_0 \\ a_1 \\ a_2 \\ a_3 \end{bmatrix} = \begin{bmatrix} 2 \\ \pi \\ \pi^2 - 4 \\ \pi(\pi^2 - 6) \end{bmatrix}$$

Applying any of our linear equation solvers (*LU* factorization or Gauss elimination) we obtain the solution

$$a_0 = -0.0505, \ a_1 = 1.3122, \ a_2 = -0.4177, \ a_3 = 0$$

so that the required least squares approximation is

$$\sin x \approx -0.0505 + 1.3122x - 0.4177x^2$$

which, for example, yields the approximations

$$\sin \pi/2 \approx 0.9802, \ \text{and} \ \sin \pi/4 \approx 0.7225$$

This solution can be easily obtained using the following MATLAB commands:

```
» c=pi.^(1:7)./(1:7);
» b=[2;pi;pi^2-4;pi*(pi^2-6)];
» for i=1:4
      for j=1:4
      A(i,j)=c(i+j-1);
   end
end
» a=A\b;
» p=fliplr(a');
```

which produces the coefficient vector to use with MATLAB's polyval function.

We can then plot the approximating polynomial on the same axes as the sine function:

```
» x=linspace(0,pi,101);
» plot(x,sin(x),x,polyval(p,x),'k')
```

This plot is shown as Figure 7.1(a). In Figure 7.1(b) we show the error function. This plot is obtained using

```
» plot(x,sin(x)-polyval(p,x))
```

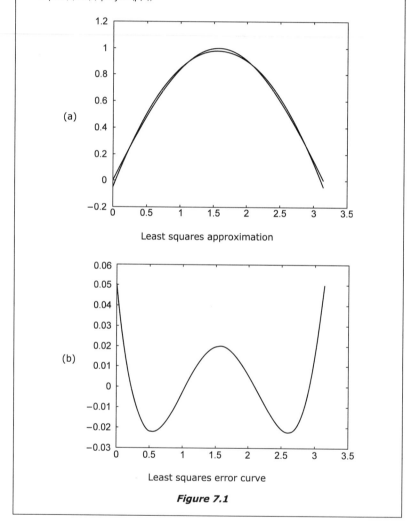

(a)

Least squares approximation

(b)

Least squares error curve

Figure 7.1

Discrete least squares problems can be solved in a similar manner.

Example 12 **Find the least squares cubic approximation to the discrete data** $(x_k, \sin(x_k))$ **where** $x_k = k\pi/8$ **for** $k = 0, 1, \ldots, 8$

The corresponding MATLAB commands to those used for Example 11 are:

```
» for k=0:6
    c(k+1)=sum(xdat.^k);
end
» for k=0:3
    b(k+1)=sum(xdat.^k.*sin(xdat));
end
» for i=1:4
    for j=1:4
        A(i,j)=c(i+j-1);
    end
end
» a=A\b;
```

These yield the coefficients of the cubic approximating polynomial which is

$$-0.0209 + 1.2648x - 0.4026x^2$$

which again has cubic coefficient 0. In Figure 7.2 we show the error curve for this approximation obtained using

```
» pd=fliplr(a');
» plot(x,sin(x)-polyval(pd,x))
```

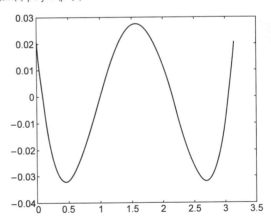

Figure 7.2 Discrete least squares error

For this particular case, the discrete least squares approximation has a smaller maximum error than the continuous one.

The MATLAB function polyfit performs essentially these operations to compute discrete least squares polynomial fitting.

One drawback of the approach described so far is that the system of linear equations tends to be ill-conditioned, especially for high-degree approximating polynomials. To see that this can happen, consider the continuous least squares polynomial approximation over $[0, 1]$. The coefficients of the linear system will then be

$$h_{ij} = c(i + j - 1) = \int_0^1 x^{i+j-2} dx = \frac{1}{i + j - 1}$$

which is to say the coefficient matrix will be the Hilbert matrix of appropriate size. We have already commented on the ill-conditioned nature of Hilbert matrices, and seen that, as the dimension increases, solutions of such systems become increasingly erroneous.

Even for discrete least squares, the coefficient matrix tends to be ill-conditioned. Alternative methods are desirable in order to obtain high-accuracy solutions. These alternatives are based on using a different basis for the representation of our approximating polynomial. The idea is to choose a basis for which the linear system will be well-conditioned and may have some sparseness that can be exploited also. The ideal level of sparsity would be for the matrix to be diagonal. We can achieve this by using *orthogonal polynomials* as our basis functions.

Suppose that $\phi_0, \phi_1, \ldots, \phi_M$ are polynomials of degrees $0, 1, \ldots, M$ respectively. These polynomials form a basis for the space of polynomials of degree $\leq M$. Therefore our approximating polynomial p can be written as a linear combination of these

$$p(x) = a_0 \phi_0(x) + a_1 \phi_1(x) + \cdots + a_M \phi_M(x) \tag{7.33}$$

and we now wish to minimize

$$F(a_0, a_1, \ldots, a_M) = \int_a^b [f(x) - (a_0 \phi_0(x) + a_1 \phi_1(x) + \cdots + a_M \phi_M(x))]^2 dx$$

In just the same manner as before, we can set all partial derivatives of F to zero to obtain a system of linear equations for the unknown coefficients. We obtain

$$\begin{bmatrix} c_{00} & c_{01} & c_{02} & \cdots & c_{0M} \\ c_{10} & c_{11} & c_{12} & \cdots & c_{1M} \\ c_{20} & c_{21} & c_{22} & \cdots & c_{2M} \\ \cdots & & \cdots & & \cdots \\ c_{M0} & c_{M1} & c_{M2} & \cdots & c_{MM} \end{bmatrix} \begin{bmatrix} a_0 \\ a_1 \\ a_2 \\ \vdots \\ a_M \end{bmatrix} = \begin{bmatrix} b_0 \\ b_1 \\ b_2 \\ \vdots \\ b_M \end{bmatrix} \tag{7.34}$$

where

$$c_{ij} = c_{ji} = \int_a^b \phi_i(x)\phi_j(x)dx$$

$$b_i = \int_a^b f(x)\phi_i(x)dx$$

If we can find a set of polynomials $\phi_0, \phi_1, \ldots$ such that each ϕ_i is of degree i and such that

$$c_{ij} = \int_a^b \phi_i(x)\phi_j(x)dx = 0 \tag{7.35}$$

whenever $i \neq j$, then the system (7.34) reduces to a diagonal system

$$\begin{bmatrix} c_{00} & 0 & 0 & \cdots & 0 \\ 0 & c_{11} & 0 & \cdots & 0 \\ 0 & 0 & c_{22} & \cdots & 0 \\ \cdots & & \cdots & & \cdots \\ 0 & 0 & \cdots & 0 & c_{MM} \end{bmatrix} \begin{bmatrix} a_0 \\ a_1 \\ a_2 \\ \vdots \\ a_M \end{bmatrix} = \begin{bmatrix} b_0 \\ b_1 \\ b_2 \\ \vdots \\ b_M \end{bmatrix} \tag{7.36}$$

which has the particularly simple solution

$$a_j = \frac{b_j}{c_{jj}} = \frac{\int_a^b f(x)\phi_i(x)dx}{\int_a^b [\phi_i(x)]^2 dx} \tag{7.37}$$

Such polynomials are known as *orthogonal polynomials* over $[a, b]$. The members of such a system of orthogonal polynomials will depend on the interval $[a, b]$. We should observe that there are many classes of orthogonal polynomials and other orthogonal functions which can be used for approximation in different circumstances. This provides just a taste of this important area of approximation theory.

One of the fundamental systems of orthogonal functions are the *Legendre polynomials* $P_n(x)$ which are orthogonal on the interval $[-1, 1]$. Multiplying orthogonal functions by scalar constants does not affect their orthogonality and so some normalization is needed. For the Legendre polynomials one common normalization is to set $P_n(1) = 1$ for each n. With this normalization the first few are

$$\begin{aligned} P_0(x) &= 1, \quad P_1(x) = 1 \\ P_2(x) &= (3x^2 - 1)/2 \\ P_3(x) &= (5x^3 - 3x)/2 \\ P_4(x) &= (35x^4 - 30x^2 + 3)/8 \end{aligned} \tag{7.38}$$

Example 13 **Find the first three orthogonal polynomials on the interval** $[0, \pi]$ **and use them to find the least squares quadratic approximation to** $\sin x$ **on this interval**

We shall normalize the polynomials to have leading coefficient unity. The first one, ϕ_0 has degree zero and leading coefficient 1. It follows that

$$\phi_0(x) = 1$$

Next $\phi_1(x)$ must have the form $\phi_1(x) = x + a$ for some constant a, and it must be orthogonal to ϕ_0. That is

$$\int_0^\pi \phi_1(x)\phi_0(x)dx = \int_0^\pi (x+a)(1)dx = 0$$

It follows that $\pi^2/2 + a\pi = 0$ so that $a = -\pi/2$ and

$$\phi_1(x) = x - \frac{\pi}{2}$$

We can write $\phi_2(x) = x^2 + bx + c$ and must choose b, c so that it is orthogonal to both ϕ_0 and ϕ_1. It is equivalent to force ϕ_2 to be orthogonal to both $1, x$. These conditions are

$$\int_0^\pi \left(x^2 + bx + c\right)dx = \frac{\pi^3}{3} + \frac{b\pi^2}{2} + c\pi = 0$$

$$\int_0^\pi x\left(x^2 + bx + c\right)dx = \frac{\pi^4}{4} + \frac{b\pi^3}{3} + \frac{c\pi^2}{2} = 0$$

which yield the solution $b = -\pi$ and $c = \pi^2/6$ so that

$$\phi_2(x) = x^2 - \pi x + \pi^2/6$$

The coefficients of the system (7.36) are then

$$c_{00} = \pi, \quad c_{11} = \frac{\pi^3}{12}, \quad c_{22} = \frac{\pi^5}{180}$$

$$b_0 = 2, \quad b_1 = 0, \quad b_2 = \frac{\pi^2}{3} - 4$$

giving the coefficients of the approximation

$$a_0 = \frac{2}{\pi}, \quad a_1 = \frac{0}{\pi^3/12}, \quad a_2 = \frac{(\pi^2 - 12)/3}{\pi^5/180}$$

The approximation is therefore

$$\sin x \approx \frac{2}{\pi} + \frac{60\pi^2 - 720}{\pi^5}\left(x^2 - \pi x + \pi^2/6\right)$$

Rearranging this in terms of the conventional basis, we get

$$\sin x \approx -0.0505 + 1.3122x - 0.4177x^2$$

which is the same approximation as we obtained in Example 11. (Note that this last rearrangement is *only* performed here for the purpose of comparison with our earlier example.)

One immediate benefit of an orthogonal basis for least squares approximation is that because the linear system becomes diagonal additional terms can be added without minimal effort. The coefficients already found remain valid for the higher-degree approximation and only the integration and division is needed to get the next coefficient.

This section serves to give just a brief introduction to an important area of numerical analysis, and to demonstrate the benefit of doing a little mathematics before starting the solution process. We see here that a poorly conditioned and difficult problem has been reduced, by using the ideas of orthogonality, to the simplest of linear algebraic problems, the solution of a diagonal system of equations. There are additional computational advantages. Not only can further coefficients and terms be added without losing any of the work already completed, but the series expansion that would result from continuing the process is usually much more rapidly convergent than a simple power series expansion. The overall result is that we can usually obtain greater accuracy for less effort.

There is nothing inherently special about using polynomials for least squares approximations. There are many situations, where other functions are appropriate. One possibility is the use of splines in place of polynomials. For periodic functions, such as waves, including square or triangular waves, it is often advantageous to use periodic functions as our basis. On the interval $[-\pi, \pi]$, the functions $\sin(mx)$ and $\cos(nx)$ form an orthogonal family. The least squares approximation using these functions is the *Fourier series*. The coefficients of this series represent the Fourier transform of the original function.

These trigonometric functions also have a discrete orthogonality property which leads to the *discrete Fourier transform* – and in special cases to the famous *Fast Fourier Transform*, or *FFT*. The interested reader is referred to more advanced texts for a description of these topics.

Other classes of functions which have become important in this context in recent years are *wavelets*. There are many wavelet bases which have uses in image compression and other areas of approximation. The basic advantage of these over Fourier series is for representing signals which have a shorter span. The simplest wavelet functions are the *Haar wavelets* which are based on one cycle of a simple square-wave function

$$\psi(x) = \begin{cases} 1 \text{ if } 0 \leq x \leq 1/2 \\ -1 \text{ if } 1/2 < x \leq 1 \\ 0 \text{ otherwise} \end{cases}$$

This function, with translation and scaling, can be used to form an orthogonal basis for approximating functions with *compact support*, which is to say the function is zero outside a closed bounded interval. Again, details of wavelet transforms are the subject of many books in their own right. We do not pursue them here.

Exercises: Section 7.5

1   Find the least squares cubic approximation to $\cos \pi x$ on $[-1, 1]$.

2   Verify that $P_0, P_1, \ldots, P_4$ given by (7.38) are orthogonal over $[-1, 1]$.

3   Evaluate $\int_{-1}^{1} x^k \cos \pi x dx$ for $k = 0, 1, \ldots, 4$. (Use a Computer Algebra System, such as Maple, or the MATLAB Symbolic toolbox if you have them.) Use these to obtain the least squares quartic (degree 4) approximation to $\cos \pi x$ on $[-1, 1]$. Plot this function, and your answer to Exercise 1, and $\cos \pi x$ on the same set of axes. Also compare the error curves.

4   Compute the discrete least squares quadratic approximation to $\cos \pi x$ on $[-1, 1]$ using 21 equally spaced data points $x = -1 : 0.1 : 1$.

5   Repeat Exercise 4 for the quartic approximation. Compare the error curves.

6   Find the least squares quartic approximations to $|x|$ on $[-1, 1]$ using

  (a)   the discrete metric with data points at $-1 : 1/5 : 1$, and
  (b)   the continuous least squares metric.

  Plot the two approximations on the same set of axes.

7   Find the orthogonal polynomials of degree up to 4, each with leading coefficient 1, on $[0, 1]$.

8   Use the results of Exercise 7 to obtain the fourth degree least squares approximation to $\arctan x$ on $[0, 1]$. Plot the error curve for this approximation.

7.6   Eigenvalues

In this section, we introduce some of the fundamental methods for computing *eigenvalues* of matrices. Recall that λ is an *eigenvalue* of a matrix A if there exists a nonzero vector, an associated *eigenvector*, $\mathbf{v}$ such that

$$A\mathbf{v} = \lambda \mathbf{v} \tag{7.39}$$

We shall refer to $(\lambda, \mathbf{v})$ as an eigen *pair*. Immediate consequences of this definition are that if λ is an eigenvalue of A then the system of equations

$$(A - \lambda I)\mathbf{x} = \mathbf{0}$$

has nontrivial solutions and, therefore,

$$\det(A - \lambda I) = 0 \qquad (7.40)$$

Equation (7.40) represents a polynomial equation in λ which, in principle could be used to obtain the eigenvalues of A. This equation, the *characteristic equation* of A is almost *never* a sensible way to go about solving the eigenvalue problem. (It is probably suitable for 2×2 matrices!)

Consideration of the characteristic equation shows that eigenvalues of real matrices can be real or complex, and can be repeated. Thus we talk about the *algebraic multiplicity* of an eigenvalue in just the same way as we do for roots of a polynomial equation. The *geometric multiplicity* of an eigenvalue is the dimension of the vector space of its associated eigenvectors. For an $n \times n$ matrix, the sum of the algebraic multiplicities of its eigenvalues is n. If the sum of the geometric multiplicities is less than n, the matrix is called defective.

We recall three other basic facts which are useful in determining eigenvalues:

1 If A is real and symmetric, then its eigenvalues are real.
2 The trace of A, $\sum a_{ii}$ equals the sum of its eigenvalues.
3 The determinant of A, $\det A$ equals the product of its eigenvalues.

We have mentioned the term *ill-conditioned* in discussing linear systems of equations. A matrix is called ill-conditioned if its *condition number* is large. Strictly speaking a matrix has many condition numbers, and 'large' is not itself well-defined in this context. The most commonly used condition number for a matrix is defined by

$$\kappa(A) = \frac{|\lambda_{\max}|}{|\lambda_{\min}|} \qquad (7.41)$$

where $\lambda_{\max}$ and $\lambda_{\min}$ are the (absolute) largest and smallest eigenvalues of A. To get an idea of what 'large' means, the condition number of the 6×6 Hilbert matrix is around 15×10^6 while that for the well-conditioned matrix

$$\begin{bmatrix} 1 & 2 & 3 & 4 & 5 & 6 \\ 2 & 2 & 3 & 4 & 5 & 6 \\ 3 & 3 & 3 & 4 & 5 & 6 \\ 4 & 4 & 4 & 4 & 5 & 6 \\ 5 & 5 & 5 & 5 & 5 & 6 \\ 6 & 6 & 6 & 6 & 6 & 6 \end{bmatrix}$$

is approximately 100.

We shall concentrate here on one method, and some variations of it, for finding eigenvalues. This method, the *power method* has the added benefit that an associated eigenvector is computed along with the eigenvalue.

Guide to Scientific Computing

In its basic form the power method is a technique which will provide the *largest* (in absolute value) eigenvalue of a square matrix A. The method is simple to describe. It is based on the following observation. If $(\lambda, \mathbf{v})$ form an eigen pair, then

$$\frac{\mathbf{v}'A\mathbf{v}}{\mathbf{v}'\mathbf{v}} = \frac{\mathbf{v}'(\lambda\mathbf{v})}{\mathbf{v}'\mathbf{v}} = \lambda\frac{||\mathbf{v}||^2}{||\mathbf{v}||^2} = \lambda \tag{7.42}$$

The ratio $\dfrac{\mathbf{v}'A\mathbf{v}}{\mathbf{v}'\mathbf{v}}$ is called the *Rayleigh quotient*. We shall use the standard Euclidean norm $||\mathbf{v}||^2 = \mathbf{v} \cdot \mathbf{v}$ throughout our description though other vector norms could be used.

Algorithm 6   Basic power method using Rayleigh quotient

Input   Square matrix A, and initial nonzero vector $\mathbf{x}_0$; tolerance ε

Initialize $k = 0$

Repeat $\mathbf{v} := A\mathbf{x}_k$

$$\lambda_{k+1} := \frac{\mathbf{v}'\mathbf{x}_k}{||\mathbf{x}_k||^2}$$

$\mathbf{x}_{k+1} := \mathbf{v}/||\mathbf{v}||$

$k = k + 1$

until $|\lambda_k - \lambda_{k-1}| < \varepsilon$

Output   Eigen pair $(\lambda_k, \mathbf{x}_k)$

The idea is that each time the vector is multiplied by the matrix A, the scale factor between $\mathbf{x}_k$ and $\mathbf{v}$ will get closer to the dominant eigenvalue $\lambda_{\max}$. This convergence is fairly easily proved provided there is a dominant eigenvalue, and provided we do not make a particularly unfortunate choice of $\mathbf{x}_0$. The details of the proof are omitted. The basis of the proof is that the vector $\mathbf{x}_0$ can be written as a linear combination of eigenvectors corresponding to the different eigenvalues. As this vector is multiplied by increasing powers of the matrix A, the contribution corresponding to the largest eigenvalue will dominate and so the Rayleigh quotient will converge to this largest eigenvalue.

Note that the renormalization step $\mathbf{x}_{k+1} := \mathbf{v}/||\mathbf{v}||$ in Algorithm 6 is only needed to counteract the potential for enormous growth in the elements of the vectors generated. Simply setting $\mathbf{x}_{k+1} := A\mathbf{x}_k$ would lead to the same eigenvalue provided the vector components do not overflow before convergence is attained. In this version of the algorithm, we have

$$\mathbf{x}_k = A^k\mathbf{x}_0 \tag{7.43}$$

which explains the origin of the name 'power method' for this algorithm.

Variants of the algorithm use different ratios in place of the Rayleigh quotient. Other possibilities are the (absolute) largest components of the vector, the (absolute) sum of the vector components, the first elements of the vectors and so on.

For vectors of moderate dimension, the Euclidean norm is easily enough computed that there is no great advantage in using these cheaper alternatives.

We note that the existence of a dominant eigenvalue for a real matrix necessarily implies that this dominant eigenvalue is *real*. The power method and its variants are therefore designed for finding real eigenvalues of real matrices or complex eigenvalues of complex matrices. They are not suited to the (common) situation of real matrices with complex eigenvalues.

Example 14   Find the largest eigenvalue of the matrix

$$A = \begin{bmatrix} 1 & 2 & 3 & 4 & 5 & 6 \\ 2 & 2 & 3 & 4 & 5 & 6 \\ 3 & 3 & 3 & 4 & 5 & 6 \\ 4 & 4 & 4 & 4 & 5 & 6 \\ 5 & 5 & 5 & 5 & 5 & 6 \\ 6 & 6 & 6 & 6 & 6 & 6 \end{bmatrix}$$

using the power method

With the arbitrary initial guess $\mathbf{x}_0 = [1, 1, 1, 1, 1, 1]'$, the results of the first few iterations are

λ	*Approximate associated eigenvector*
26.8333	$[0.3135, 0.3284, 0.3583, 0.4031, 0.4628, 0.5374]'$
27.6991	$[0.3318, 0.3431, 0.3663, 0.4024, 0.4530, 0.5204]'$
27.7224	$[0.3288, 0.3407, 0.3651, 0.4026, 0.4547, 0.5231]'$
27.7230	$[0.3293, 0.3411, 0.3653, 0.4026, 0.4544, 0.5227]'$
27.7230	$[0.3292, 0.3411, 0.3652, 0.4026, 0.4545, 0.5228]'$

We see that the eigenvalue estimates have already settled to four decimals and the components of the associated eigenvector are also converging fairly rapidly in this case.

The power method appears to be reasonably efficient for finding the dominant eigenvalue of a matrix. What can we do to find the others?

The next easiest eigenvalue to obtain is the (absolute) smallest one. The reason for this is that for a nonsingular matrix A, eigenvalues of A^{-1} are the reciprocals of those of A. This is easy to see since if $(\lambda, \mathbf{v})$ are an eigen pair of A, then we have

$$A\mathbf{v} = \lambda\mathbf{v}$$

and premultiplying by A^{-1} we get $\mathbf{v} = \lambda A^{-1}\mathbf{v}$, or

$$\frac{1}{\lambda}\mathbf{v} = A^{-1}\mathbf{v}$$

It follows that the smallest eigenvalue of A is the reciprocal of the largest eigenvalue of A^{-1}. This largest eigenvalue of A^{-1} could be computed using the power method – except we do not have A^{-1}. This technique, known as *inverse iteration*, can be implemented by taking advantage of the LU factorization of A.

At each iteration, we wish to find $\mathbf{v} = A^{-1}\mathbf{x}_k$, which is equivalent to solving the system

$$A\mathbf{v} = \mathbf{x}_k$$

and, if we know the LU factors of A, we can use these and forward and back substitution to achieve this. The algorithm for inverse iteration can therefore be written as follows.

Algorithm 7   Inverse iteration using LU factorization

Input   Matrix A, initial nonzero vector $\mathbf{x}_0$; tolerance ε
Initialize   Find **LU** factorization $LU = A$
            Set $k = 0$
Repeat   Solve $L\mathbf{w} = \mathbf{x}_k$
            Solve $U\mathbf{v} = \mathbf{w}$
            $\lambda_k := \dfrac{\mathbf{v}'\mathbf{x}_k}{\|\mathbf{x}_k\|^2}$
            $\mathbf{x}_{k+1} := \mathbf{v}/\|\mathbf{v}\|$
            $k = k + 1$
until $|\lambda_k - \lambda_{k-1}| < \varepsilon$
Output   Eigen pair $(1/\lambda_k, \mathbf{x}_k)$

Here we see a great benefit from using the LU factorization since several iterations are likely to be needed to obtain the eigenvalue with high accuracy.

Example 15   Use inverse iteration to compute the smallest eigenvalue of the matrix A used in Example 14

Using the same starting vector as in Example 14 the first few estimates of the largest eigenvalue of the inverse of A are: 0.0278, –0.8333, –2.5055, –3.0017 which are settling slowly. After another 22 iterations we have the converged value –3.7285 for the largest eigenvalue of A^{-1} so that the smallest eigenvalue of A is –0.2682 with the approximate associated eigenvector $[0.1487, -0.4074, 0.5587, -0.5600, 0.4076, -0.1413]'$.

Testing the residual of these estimates by computing $A\mathbf{v} - \lambda\mathbf{v}$ we get a maximum component of approximately 3×10^{-4} which is consistent with the accuracy to which we have computed the eigenvalue itself.

We can obviously improve the performance by choosing a better initial vector. In Example 14, we saw that the eigenvector corresponding to the largest eigenvalue has components which are all quite similar in magnitude and of the same sign. We know that eigenvectors corresponding to distinct eigenvalues are orthogonal so we might reasonably try an initial guess in which the components of the vector alternate in sign such as $[1, -1, 1, -1, 1, -1]'$. With this initial guess, just 16 iterations are sufficient to obtain $\lambda_{min} = -0.26819847819867$ with the tolerance set at 10^{-8}. (The actual error in λ_{min} is around 9×10^{-11}.)

So far we have been able to use the power method to find the largest and smallest eigenvalues of a matrix. What about intermediate ones? There is another variation on the basic idea which can be helpful here, too.

First it will be helpful to get some idea of where the other eigenvalues may be. There is a famous theorem which can often help.

Theorem 1 (Gerschgorin's Theorem)
Every eigenvalue of an $n \times n$ matrix A lies in one of the complex disks centered on the diagonal elements of A with radius equal to the sum of the off-diagonal elements in the corresponding row. These are the disks

$$|z - a_{ii}| \leq \sum_{j \neq i} |a_{ij}|$$

Moreover, if any collection of m of these disks is disjoint from the rest, then exactly m eigenvalues (counting multiplicities) lie in the union of these disks.

Example 16   Apply Gerschgorin's theorem to the matrix of Examples 14 and 15

The diagonal entries and the absolute sums of their off-diagonal row-elements give us the following set of centers and radii:

Center	1	2	3	4	5	6
Radius	20	20	21	23	26	30

These disks are plotted in Figure 7.3.

In this case Gerschgorin's Theorem gives us little new information since all the disks are contained in the largest one. The matrix A is symmetric and so all its eigenvalues are real. Therefore the theorem only tells us that all eigenvalues lie in $[-24, 36]$. Since we already know that $\lambda_{max} = 27.7230$ and $\lambda_{min} = -0.2682$ to four decimals, we can reduce this interval to $[-24, 27.7230]$ but this does not help much in locating the other eigenvalues.

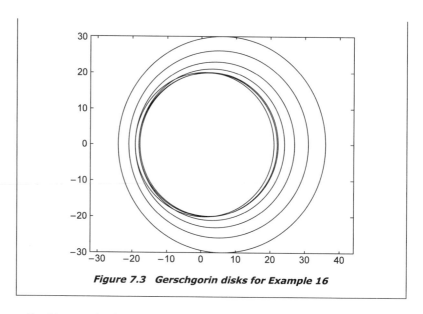

Figure 7.3 Gerschgorin disks for Example 16

For this example, Gerschgorin's theorem is not much help, but that is not always the case, of course. The matrix

$$B = \begin{bmatrix} 3 & 0 & 1 & -1 \\ 0 & 6 & 0 & 1 \\ 1 & 0 & -1 & 0 \\ -1 & 1 & 0 & 10 \end{bmatrix} \tag{7.44}$$

is symmetric and so has real eigenvalues. Gerschgorin's theorem tells us that these lie in the union of the intervals $[1,5]$, $[5,7]$, $[-2,0]$, and $[8,12]$. Furthermore since the last two of these are each disjoint from the others, we can conclude that one eigenvalue lies in each of the intervals $[-2,0]$, and $[8,12]$ while the remaining two lie in $[1,5] \cup [5,7] = [1,7]$. The smallest and largest are -1.2409 and 10.3664 to four decimal places, respectively. There are two eigenvalues in $[1,7]$ still to be found. If we can locate the one closest to the mid-point of this interval, then we could use the trace to obtain a very good approximation to the remaining one.

How then could we compute the eigenvalue closest to a particular value? Specifically suppose we seek the eigenvalue of a matrix A closest to some fixed μ.

We observe first that if λ is an eigenvalue of A with associated eigenvector $\mathbf{v}$, then $\lambda - \mu$ is an eigenvalue of $A - \mu I$ with the same eigenvector $\mathbf{v}$ since

$$(A - \mu I)\mathbf{v} = A\mathbf{v} - \mu\mathbf{v} = \lambda\mathbf{v} - \mu\mathbf{v} = (\lambda - \mu)\mathbf{v}$$

The smallest eigenvalue of $A - \mu I$ can be found using inverse iteration. Adding μ will then yield the desired eigenvalue of A closest to μ.

This technique is called *origin shifting* since we 'shift the origin' of the number line on which our eigenvalues lie.

Example 17 Find the eigenvalue of the matrix B given by (7.44) closest to 4

Applying inverse iteration to the matrix

$$C = B - 4I = \begin{bmatrix} -1 & 0 & 1 & -1 \\ 0 & 2 & 0 & 1 \\ 1 & 0 & -5 & 0 \\ -1 & 1 & 0 & 6 \end{bmatrix}$$

gives the smallest eigenvalue as -0.907967 to six decimals. Adding 4 gives the closest eigenvalue of B to 4 as 3.092033.

The trace of B is $3 + 6 - 1 + 10 = 18$ and this must be the sum of the eigenvalues. We therefore conclude that the remaining eigenvalue must be very close to

$$18 - (-1.2409) - 3.0920 - 10.3664 = 5.7825$$

If this eigenvalue was sought to greater accuracy, we could use inverse iteration with an origin shift of, say, 6.

Although we have not often reported them, all the techniques based on the power method also provide the corresponding eigenvectors.

What about the remaining eigenvalues of our original example matrix from Example 14? Gerschgorin's theorem provided no new information. The matrix was

$$A = \begin{bmatrix} 1 & 2 & 3 & 4 & 5 & 6 \\ 2 & 2 & 3 & 4 & 5 & 6 \\ 3 & 3 & 3 & 4 & 5 & 6 \\ 4 & 4 & 4 & 4 & 5 & 6 \\ 5 & 5 & 5 & 5 & 5 & 6 \\ 6 & 6 & 6 & 6 & 6 & 6 \end{bmatrix}$$

and we have largest and smallest eigenvalues $\lambda_{max} = 27.7230$ and $\lambda_{min} = -0.2682$ to four decimals.

Example 18   Find the remaining eigenvalues of the matrix A above

The sum of the eigenvalues is the trace of the matrix, 21, and their product is $\det A = -6$. Since

$$\lambda_{max} + \lambda_{min} = 27.7230 - 0.2682 = 27.4548$$

the sum of the remaining values must be approximately -6.5. Their product must be close to $-6/(27.7230)(-0.2682) = 0.807$. Given the size of the product it is reasonable to suppose there is one close to -5 or -6 with three smaller negative ones.

Using an origin shift of -5 and inverse iteration on $A - (-5)I$ we get one eigenvalue at -4.5729.

The final three have a sum close to -2 and a product close to $0.8/(-4.6) = -0.174$. Assuming all are negative, then one close to -1 and two small ones looks likely. Using an origin shift of -1, we get another eigenvalue of A at -1.0406.

Finally, we need two more eigenvalues with a sum close to -1 and a product around 0.17. Using an origin shift of -0.5 we get the fifth eigenvalue -0.5066. The trace now gives us the final eigenvalue as being very close to

$$21 - 27.7230 - (-4.5729) - (-1.0406) - (-0.5066) - (-0.2682)$$
$$= -0.3347$$

What we have seen in this last example is that for a reasonably small matrix, which we can examine, then the combination of the power method with inverse iteration and origin shifts *can* be used to obtain the full eigen structure of a matrix all of whose eigenvalues are real. This set of hypotheses are very restrictive. If a matrix is generated by the overall numerical process, then we may well not have it available for this type of scrutiny. We cannot usually know in advance that all eigenvalues are real. The possible sign patterns of the 'missing' eigenvalues might be much more difficult to guess than in this example.

There is a clear need therefore for more automatic algorithms for obtaining all the eigenvalues of a matrix. Such methods do indeed exist but are *much* more difficult to explain and analyze. They are typically the subject of advanced courses dedicated to numerical linear algebra. This section, as promised at the outset, serves only to introduce the subject and perhaps whet your appetite for such a course later in your career. MATLAB has advanced algorithms for eigenvalues and the more general singular values among its powerful suite of linear algebra functions.

Exercises: Section 7.6

Consider the matrix

$$A = \begin{bmatrix} 6 & 6 & 6 & 6 & 6 & 6 \\ 5 & 5 & 5 & 5 & 5 & 6 \\ 4 & 4 & 4 & 4 & 5 & 6 \\ 3 & 3 & 3 & 4 & 5 & 6 \\ 2 & 2 & 3 & 4 & 5 & 6 \\ 1 & 2 & 3 & 4 & 5 & 6 \end{bmatrix}$$

which has entirely real eigenvalues. Find these eigenvalues each accurate to 5 decimal places.

1 Find the largest eigenvalue of A using the power method.
2 Find the smallest eigenvalue of A using inverse iteration. (Write a program for inverse iteration as you will be using it in the subsequent exercises.)

3 Given that there is positive eigenvalue close to 5, use inverse iteration with an origin shift to locate this eigenvalue.

4 Repeat Exercise 3 for the eigenvalue close to -1.

5 Use the trace and determinant to find a suitable origin shift and then another eigenvalue.

6 Use the trace to estimate the final eigenvalue. Use a suitable origin shift to compute this last one and to provide an accuracy check.

7.7 MATLAB's linear algebra functions

7.7.1 Linear equations

The fundamental MATLAB operator for linear equation solving is the 'backslash' \ operator which can be interpreted as 'divided into'.

The \ operator We have already seen the use of this operator in a small number of places. Essentially it is the command to solve a system of linear equations. The syntax is simple. To solve a system

$$A\mathbf{x} = \mathbf{b}$$

we can just set

```
» x=A\b;
```

For a generic square system the algorithm used is LU factorization followed by forward and back substitution. If the system is triangular in the first place then just the appropriate forward or back substitution is used. In the case where the system is not square the system is solved in a least squares sense using the QR factorization (see below).

LU factorization Again, we have already discussed this algorithm and used the MATLAB routine in a number of places. The syntax is straightforward:

```
[L,U]=lu(A);
```

generates the LU factors of A using partial pivoting so that L need not be strictly lower triangular as the row interchanges will be incorporated.

```
[L,U,P]=lu(A);
```

uses exactly the same algorithm but returns the three matrices such that

$$LU = PA$$

where L, U are genuine lower and upper triangular matrices and P is the permutation matrix which reflects the order in which the rows were used. To solve the system $A\mathbf{x} = \mathbf{b}$, we would therefore solve $LU\mathbf{x} = P\mathbf{b}$ by setting

```
» x=U\(L\ (P*b));
```

QR factorization   This, numerically more stable, factorization of a matrix A into an orthonormal matrix Q and an upper triangular matrix R uses a more advanced algorithm than any we have studied in this chapter. An *orthonormal matrix* is one whose columns are mutually *ortho*gonal and are *normal*ized to each have norm 1. This factorization can be applied to square or rectangular matrices. If A is $m \times n$ then Q is an $m \times m$ orthonormal matrix and R is upper 'triangular' $m \times n$.

The advantage of an orthonormal matrix is that its transpose is its inverse, so that solving $Q\mathbf{y} = \mathbf{b}$ is equivalent to simply setting $\mathbf{y} = Q'\mathbf{b}$. The solution of a linear system $A\mathbf{x} = \mathbf{b}$ can therefore be achieved by setting

```
» [Q,R]=qr(A);
» x=R \(Q'*b);
```

where again the syntax is straightforward.

7.7.2   Linear least squares

polyfit   The polyfit function obtains the discrete least squares fit to a set of data. It does *not* form the system of normal equations but rather uses the QR factorization just described to solve the overdetermined linear system in a least squares sense. This linear system is formed as a rectangular Vandermonde system similar to that which could be used for Lagrange interpolation.

If the data points are x_k $(k = 0, 1, \ldots, n)$ (with corresponding function values f_k) and we wish to fit a polynomial of degree $m < n$, the linear system becomes

$$a_0 + a_1 x_k + a_2 x_k^2 + \cdots + a_m x_k^m = f_k$$

for $k = 0, 1, \ldots, n$. There are therefore n equations in the m unknown coefficients. To solve such a system in the least squares sense means to find values $a_0, a_1, \ldots, a_m$ such that the sum of the squared residuals

$$\sum_{k=0}^{n} \left| f_k - \left(a_0 + a_1 x_k + a_2 x_k^2 + \cdots + a_m x_k^m \right) \right|^2$$

is minimized.

This of course is precisely the same minimization that leads to the normal equations. The QR method however provides a more efficient and more robust method for solving this problem.

Note: MATLAB returns the vector of coefficients beginning with the highest powers. This is consistent with all the MATLAB polynomial functions. For example with the data defined by

```
» x=-1:0.1:1; y=cos(pi*x);
```

the command

```
» p=polyfit(x,y,3)
```

returns the coefficients of the least squares cubic approximation as

p =
  −0.0000 −2.1039 0.0000 0.7238

which represents the 'cubic'

$$-0x^3 - 2.1039x^2 + 0x + 0.7238$$

7.7.3 Eigenvalues

eig   The MATLAB command eig is the basic eigenvalue and eigenvector routine. It is built on an extension of the QR factorization algorithm for linear systems and returns all eigenvalues (and eigenvectors) of a square matrix.

The command

» d=eig(A);

returns a vector containing all the eigenvalues of the matrix A. If the eigenvectors are also wanted, the syntax

[V,D]=eig(A);

will return a matrix V whose columns are eigenvectors of A corresponding to the eigenvalues in the diagonal matrix D.

   The algorithm can also be used for solution of the *generalized eigenvalue problem*

$$A\mathbf{x} = \lambda B\mathbf{x}$$

svd   SVD stands for *Singular Value Decomposition*. This is a generalization of the idea of an eigenvalue decomposition of a square matrix to rectangular matrices. The idea is to factor a rectangular matrix in the form

$$A = USV'$$

where U, V are orthonormal matrices of appropriate dimensions and S is a diagonal matrix. The entries of S are the *singular values of A*. The singular values are the square-roots of the eigenvalues of the matrix $A'A$

rank   It is often desirable to compute the *rank* of a matrix. This is the number of linearly independent rows (or columns) of a matrix. It is also the number of nonzero eigenvalues of a square matrix, or the number of nonzero singular values of a rectangular matrix. MATLAB's function rank uses the singular value decomposition to determine the rank of a matrix according to this last criterion. Due to roundoff errors, a tolerance must be used to determine whether a particular singular value should be deemed to be 'zero'. MATLAB uses a default value that is based on the dimension of the matrix, the norm of the matrix (its largest singular value) and the machine unit for the computer arithmetic.

As usual the syntax is simple:

» r=rank(A);

uses the default tolerance, while

» r=rank(A,tol);

uses a user-specified tolerance tol.

7.7.4   Basic linear algebra functions

We have seen the use of standard linear algebra 'arithmetic' operations on vectors and matrices throughout. There are a small number of other functions that should be mentioned.

dot   forms the dot, or scalar, product of two vectors

$$\mathbf{a} \cdot \mathbf{b} = \sum a_k b_k$$

For example,

» a=[1,2,3]; b=[6,7,9];
» dot(a,b)

returns the answer 47.

norm   computes the Euclidean norm of a vector (or a matrix)

$$\|\mathbf{a}\| = \sqrt{\mathbf{a} \cdot \mathbf{a}}$$

The 1-norm and ∞-norm can also be obtained using norm. For example with **b** as above, we get

» norm(b)
ans =
  12.8841
» norm(b,1)
ans =
  22
» norm(b,inf)
ans =
  9

The default is the Euclidean or 2-norm.

APPENDIX A MATLAB Basics

1 MATLAB numbers and numeric formats

All numerical variables are stored in MATLAB in *double-precision floating-point* form. (In fact it is *possible* to force some variables to be of other types but not easily and this ability is not needed here.) *Floating-point* representation of numbers is essentially equivalent to the 'scientific notation' of your calculator. Specifically a (real) number x is stored in *binary* floating-point form as

$$x = \pm f * 2^E$$

where f is called the *mantissa* and E the *exponent*. The exponent is an integer (positive or negative) and the mantissa lies in the range $1 \le f < 2$. This representation is entirely *internal* to the machine and its details are not important here. (They are important when analyzing carefully the performance of numerical algorithms.)

Similar formats using the conventional *decimal* system are available for MATLAB input and output. These formats can be set in the File – Preferences menu in the MATLAB command window. They can also be changed using MATLAB commands:

```
» format short
» pi
ans =
   3.1416
» format long
» pi
ans =
   3.14159265358979
» format short e
» pi
ans =
   3.1416e+000
» 100*pi
ans =
   3.1416e+002
```

261

What does this e *mean?* The result 3.1416e+000 just means 3.1416×10^0 where three digits are allowed for the *decimal* exponent. Similarly, the final answer 3.1416e+002 means 3.1416×10^2 or, approximately, 314.16. In the same way, 3.1416e−002 means 3.1416×10^{-2} or, approximately, 0.031416

One final note on this first MATLAB session: clearly the constant π is known and is built into MATLAB to high accuracy with the name pi. You should therefore avoid using pi as a variable name in your MATLAB work!

2   Strings and printing

In more advanced applications such as symbolic computation, string manipulation is a very important topic. For our purposes, however, we shall need only very limited skills in handling strings initially. One most important use might be to include Your Name and the Course as part of your MATLAB workspace in a simple, and automatic, way.

This is easily achieved by using strings and the MATLAB print function fprintf in a special file called startup.m which will be executed automatically when you start MATLAB.

Strings can be defined in MATLAB by simply enclosing the appropriate string of characters in single quotes such as

≫ s='My Name'

results in the output

s =
My Name

More complicated strings can be printed using the MATLAB function fprintf. This is essentially a C programming command which can be used to obtain a wide range of printing specifications. Very little of its detail is needed at this stage. We use it here only to illustrate printing strings. If you want more details on advanced printing, full details are provided in your MATLAB manual. The use of fprintf to print the Name and Course information is illustrated by

≫ fprintf(' My Name \n Course \n')
  My Name
  Course
≫

where the \n is the new-line command. The final one is included to ensure that the next MATLAB prompt occurs at the beginning of the next line rather than immediately at the end of the printed string.

To make this appear automatically every time you start MATLAB, simply create a file called startup.m which has this one command in it. To create this m-file, click the 'New m-file' icon in the MATLAB window. Type the single line

fprintf(' My Name \n Course \n')

and then save this file with the name 'startup'. By default, MATLAB will save this in a folder called Work which is where you want it to be. (In older versions of MATLAB, the default location may be different, but is still acceptable.)

The function fprintf can be used for specifying the format of MATLAB's numeric output, too. However, for our purposes, the various default numerical formats are usually adequate.

3 Editing in the Command Window

Although we have not yet done very much in the MATLAB Command Window, it is worth summarizing some of the basic editing operations available there. Most of the elementary editing operations are completely standard:

- BackSpace and Delete work just as in any other Windows-based program;
- Home moves the cursor to the beginning of the current line;
- End moves the cursor to the end of the current line;
- → and ← move the cursor one place right or left;
- Ctrl → and Ctrl ← move one *word* right or left.
- Ctrl-C and Ctrl-V have their usual effect of copying and pasting.

There are some nonstandard editing shortcuts which are useful:

- Esc deletes the whole of the current line;
- Ctrl-K deletes from the cursor to the end of the current line.

The ↑ and ↓ keys have special roles:

- ↑ recalls the previous line (repeatedly up to many lines) so that earlier lines can be reproduced. This is often useful when you want to correct errors and typos in earlier code without having to retype complicated instructions – or simply to repeat an operation.
- ↓ recalls the next line. (Obviously if the current line is the last one, then there is no next line.)

These two keys are especially important because of the fact that MATLAB is interactive and each line is executed as it is completed so that you cannot go back up the Workspace to correct an earlier mistake.

4 Arithmetic operations

Arithmetic in MATLAB follows all the usual rules and uses the standard computer symbols for its arithmetic operation signs.

Thus we use

Symbol	Effect
$+$	Addition
$-$	Subtraction
$*$	Multiplication
$/$	Division
$\wedge$	Power

In our present context we shall consider these operations as *scalar* arithmetic operations, which is to say that they operate on 2 numbers in the conventional manner. MATLAB's arithmetic operations are actually much more powerful than this. We shall see just a little of this extra power later.

The conventional algebraic order of precedence between the various operations applies. That is,

(expressions in parentheses)

take precedence over

powers, $\wedge$

which take precedence over

multiplication and division, $*$, $/$

which, in turn, take precedence over

addition and subtraction, $+$, $-$

We have in fact already seen examples of some of these operations.

Assignment of values – whether direct numerical input or the result of arithmetic – is achieved with the usual $=$ sign. Therefore, within MATLAB (and other programming languages) we can legitimately write 'equations' which are mathematically impossible. For example, the assignment statement

```
>> x = x+0.1
```

has the effect of incrementing the value of the variable x by 0.1 so that if the current value of x before this statement is executed is 1.2 then its value after this is executed is 1.3. Such actions are often referred to as *overwriting* the value of x with its new value.

Similarly the MATLAB commands

```
>> i=1;
>> i=i+2
```

result in the output

```
i = 3
```

Note the use of the ; to suppress the output from the first line here.

There are some arithmetic operations which require great care. The order in which multiplication and division operations are specified is especially important.

What is the output from the following MATLAB commands?

» a=2; b=3; c=4;
» a/b*c

Here the absence of any parentheses results in MATLAB executing the two operations (which are of *equal* precedence) from left-to-right so that

• First a is divided by b,

and then

• The *result* is multiplied by c.

The result is therefore

ans =
    2.6667

Note here the default 'variable' ans is used for any arithmetic operations where the result is not assigned to a named variable.

This arithmetic is equivalent to $\frac{a}{b}c$ or as a MATLAB command

» (a/b)*c

Similarly, a/b/c yields the same result as $\frac{a/b}{c}$ or $\frac{a}{bc}$ which could (alternatively) be achieved with the MATLAB command

» a/(b*c)

Use parentheses to be sure that MATLAB does what you intend!

5 MATLAB's mathematical functions

All of the standard mathematical functions – often called the *elementary* functions – that you met in your calculus courses are available in MATLAB using their usual mathematical names. Many other functions – the *special* functions – are also included; you will most likely come across some of these in later mathematics and, more especially, engineering courses.

The elementary functions are listed in your User's Guide. This listing includes several functions which will not be familiar to you yet, and several that we shall not deal with in this book. The important functions for our purposes are:

abs (x)   Absolute value
sqrt (x)   Square root
sin (x)   Sine
cos (x)   Cosine

tan (x)	Tangent
log (x)	Natural logarithm
exp (x)	Exponential function, e^x
atan (x)	Inverse tangent, or arctan
asin (x)	Inverse sine, or arcsin
acos (x)	Inverse cosine, or arccos
cosh (x)	Hyperbolic cosine
sinh (x)	Hyperbolic sine

Note that the various trigonometric functions expect their argument to be in *radian* (or pure number) form – *NOT* in degrees which are an artificial unit based on the ancient Babylonians' belief that a year was 360 days long! For example,

» sin(pi/3)

gives the output

ans =
  0.8660

Shortly we shall see how to use these functions to generate both tables of their values and plots of their graphs. Also we shall see how to define other functions either as strings in the MATLAB command window, or more usefully, as function m-files which can be saved for repeated use.

6   Vectors

In MATLAB the word *vector* should really be interpreted simply as 'list of numbers'. Strictly it could be a list of other objects than numbers but 'list of numbers' will fit our needs for now. These *can* be used to represent physical vectors but are much more versatile than that, as we shall see.

There are two basic kinds of MATLAB vectors: *Row* and *Column* vectors. As the names suggest, a row vector stores its numbers in a long 'horizontal list' such as

$$1, 2, 3.4, 1.23, -10.3, 2.1$$

which is a row vector with 6 components. A column vector stores its numbers in a vertical list such as

$$\begin{array}{c} 1 \\ 2 \\ 3.4 \\ 1.23 \\ -10.3 \\ 2.1 \end{array}$$

which is a column vector with (the same) 6 components. In mathematical notation these arrays are usually enclosed in brackets [ ].

There are various convenient forms of these vectors, for allocating values to them and accessing the values that are stored in them. The most basic method of accessing or assigning individual components of a vector is based on using an *index*, or *subscript*, which indicates the position of the particular component in the list. The MATLAB notation for this subscript is to enclose it in parentheses ( ).

For assigning a complete vector in a single statement, we can use the [ ] notation. These two are illustrated for a row vector in the following MATLAB session:

```
» x=[1,2,1.23,3.4,-8.7,2.3]
x =
  1.0000 2.0000 1.2300 3.4000 -8.7000 2.3000
» x(2)=x(1)+2*x(3)
x =
  1.0000 3.4600 1.2300 3.4000 -8.7000 2.3000
```

The first command simply initializes the vector with the given components. The second performs arithmetic to set the second component equal to the first plus twice the third. Note that the full vector is output, and that the other components are unchanged.

In entering values for a row vector, spaces could be used in place of commas. For the corresponding column vector simply replace the commas with semi-colons.

- **All MATLAB vectors have their index or subscript begin at 1.**
  This is *NOT* something that the user can vary. For most applications this causes little difficulty but there are times when we must take special precautions in MATLAB programming to account for this.
- **To switch between column and row format for a MATLAB vector we use the *transpose* operation denoted by '.**
  This is illustrated for the row vector above by:

```
» x = x'
x =
   1.0000
   3.4600
   1.2300
   3.4000
  -8.7000
   2.3000
```

To switch back to row form: just use the transpose operator again.

MATLAB has several convenient ways of allocating values to a vector where these values fit a simple pattern.

7 MATLAB's : notation

The *colon* : has a very special and powerful role in MATLAB. Basically, it allows an easy way to specify a vector of *equally spaced* numbers.

There are 2 basic forms of the MATLAB : notation.

- **Two arguments separated by a : as in**

  » v= 2 · 4

  generates a row vector with first component 2, last one 4, and others spaced at unit intervals
- **Three arguments separated by two :s has the effect of specifying the starting value : spacing : final value**

For example, the MATLAB command

» v=−1:0.2:2;

generates the row vector

v = −1.0,−0.8,−0.6,−0.4,−0.2,0.0,0.2, 0.4, 0.6, 0.8, 1.0, 1.2, 1.4, 1.6, 1.8, 2.0.

Remember the syntax is start:step:stop. Also the step can be negative.

There are no practical restrictions on the length of such a vector (or, therefore, on the range or spacing which can be used). Of course very long vectors may have two negative effects:

- Computing times are likely to rise dramatically, and
- Output may take up so much room in the Window that you have no way of fathoming it – or printing it in reasonable time.

- **Don't forget the use of a semi-colon ; to suppress output in the Command Window**

There is another very useful aspect of MATLAB's colon notation. It can also be used to specify a *range* of subscripts for a particular operation. We see this in the following example.

» w=v(3:8)

w =

  −0.6000 −0.4000 −0.2000 0 0.2000 0.4000

Note: The index of w still starts at 1, so that, for example,

» w(3)

ans =

  −0.2000

8 linspace and logspace

MATLAB has two other commands for specifying vectors conveniently.

- **linspace** is used to specify a vector with a given number of equally spaced elements between specified start and finish points
  This needs some care with counting to get convenient spacing.
  For example:
  » x = linspace(0,1,10)
  results in the vector [0, 0.1111, 0.2222, 0.3333, 0.4444, 0.5556, 0.6667, 0.7778, 0.8889, 1.0000]. Using 10 points results in just 9 *steps*.
  » x = linspace(0,1,11)
  gives the vector [0, 0.1, 0.2, 0.3, 0.4, 0.5, 0.6, 0.7, 0.8, 0.9, 1.0].

- **logspace** has a similar effect – except that the points are spaced on a logarithmic scale

9 Tables of function values

We can use MATLAB's vectors to generate tables of function values. For example:

```
» x=linspace(0,1,11);
» y=sin(x);
» [x',y']
```

generates the output

```
ans =
  0 0
  0.1000 0.0998
  0.2000 0.1987
  0.3000 0.2955
  0.4000 0.3894
  0.5000 0.4794
  0.6000 0.5646
  0.7000 0.6442
  0.8000 0.7174
  0.9000 0.7833
  1.0000 0.8415
```

Note the use of the transpose to convert the row vectors to columns, and the separation of these 2 columns by a comma.

Note also that all the standard MATLAB functions are defined to operate on vectors of inputs in an element-by-element manner. The following example illustrates the use of the : notation and arithmetic within the argument of a function:

```
» y=sqrt(2+3*(0:0.1:1)')
y =
    1.4142
    1.5166
    1.6125
    1.7029
    1.7889
    1.8708
    1.9494
    2.0248
    2.0976
    2.1679
    2.2361
```

10   Array arithmetic

Array arithmetic allows us to perform the equivalent arithmetic operations on all the components of a vector. In most circumstances the standard arithmetic symbols achieve what is wanted and expected. However, especially for multiplication, division and powers involving MATLAB vectors (or matrices) a dot . is needed in conjunction with the usual operation sign.

These various operations are best illustrated with some simple MATLAB examples. For these examples we shall use the following data vectors a and b, and scalar (number) c.

```
» a=[1,2,3,4]; b=[5,3,0,2]; c=2;
```

Since the vectors a and b have the same size, they can be added or subtracted from one another. They can be multiplied, or divided, by a scalar, or a scalar can be added to each of their components. These operations are achieved using the following commands which produce the output shown:

```
» a+b
ans =
   6 5 3 6
» a-2*b
ans =
  -9 -4 3 0
» c*a
ans =
   2 4 6 8
```

```
» a+c
ans =
  3 4 5 6
» b/c
ans =
  2.5000 1.5000 0 1.0000
```

Mathematically the operation of division by a vector does not make sense. To achieve the corresponding componentwise operation, we use c./a. Similarly, for powers we use .^ as follows:

```
» c./a
ans =
  2.0000 1.0000 0.6667 0.5000
» a.^c % squares of components of a
ans =
  1 4 9 16
» c.^a
ans =
  2 4 8 16 % powers of c=2
```

Similarly the mathematical operation $a * b$ is not defined but we may wish to generate the vector whose components are the products of corresponding elements of a and b:

```
» a.*b
ans =
  5 6 0 8
» b.^a
ans =
  5 9 0 16
» a./b
Warning: Divide by zero.
ans =
  0.2000 0.6667 Inf 2.0000
```

Note the warning created by the division by zero in the third element of the final operation here.

11 String functions

The simplest user-defined functions in MATLAB are created as strings in the command window. If we may need to evaluate these for arrays of arguments, remember to use the 'dot operations' where necessary.

As an example, evaluate the function

$$f(x) = 2x^3 - \frac{3x}{1+x^2}$$

for $x = 0, 0.3, 0.6, \ldots, 3$.

```
» x=0:0.3:3;
» f='2*x.^3-3*x./(1+x.^2)';
» y=eval(f);
» [x',y']
ans =
  0 0
  0.3000 -0.7717
  0.6000 -0.8915
  0.9000 -0.0337
  1.2000 1.9806
  1.5000 5.3654
  1.8000 10.3904
  2.1000 17.3575
  2.4000 26.5829
  2.7000 38.3889
  3.0000 53.1000
```

The only difficulty here is remembering that each time we form products, quotients or powers of vectors, the corresponding dot operation must be used.

However there is another major drawback to this simple approach. Every time we wish to evaluate the function f, we must ensure that the argument is stored in a variable called x. This may be inconvenient if we have several functions and arguments within a computation.

For this reason, function m-files are a much superior way of creating and saving user-defined functions.

12 Function m-files

To create a new m-file, click the 'New m-file' icon in the MATLAB command window.

If the m-file is to be a function m-file, the first word of the file is function, we must also specify names for the function, its input and output. The last two of these are purely *local* variable names. These mimic the mathematical idea of defining a function

$$y = f(x)$$

but then assigning, for example, $z = f(2)$. This is just a shorthand for temporarily setting $x = 2$, evaluating the output y of the function and assigning this value to z. Function m-files work the same way.

To illustrate we again evaluate the function

$$f(x) = 2x^3 - \frac{3x}{1 + x^2}$$

for $x = 0, 0.3, 0.6, \ldots, 3$ but this time using a function m-file.

The m-file could be:

```
function y=fun1(x)
y=2*x.^3-3*x./(1+x.^2);
```

Then the commands

```
» v=(0:0.3:3)';
» fv=fun1(v);
» [v,fv]
```

generate the same table of output as before.

Note that this time, the vector v is a column vector and, consequently, so is fv. There is no need for any transpose operations in the output line.

MATLAB m-files can be *much* more involved than this, as you will see during your course, but the principles are the same.

13   Script m-files

A script m-file is a MATLAB *program*. It consists of the set of MATLAB instructions for performing a particular task. It is run by simply typing its name in the MATLAB command window. The startup.m file created earlier is a script m-file.

As well as storing complete programs, perhaps the best way to save work if you are in the middle of an assignment is to use a 'script' m-file. By copying the instructions you used into an m-file and saving it with some appropriate name, you can then recover to exactly the point at which you stopped work. Further instructions can then be added in the command window, or by editing the m-file to add the new instructions.

As a simple example of a script m-file, we could store the m-file containing the three lines

```
v=(0:0.3:3)';
fv=fun1(v);
[v,fv]
```

with the name Ex_table.m and then the command

```
» ex_table
```

will again produce the same table of function values as we generated above.

Again, obviously, script m-files can be *much* more complicated than this!

14 Plotting

MATLAB has several methods for plotting – both in 2- and 3-dimensional settings. We shall concentrate on just one of the 2-dimensional plotting functions, the most powerful, plot.

MATLAB's plot function has the ability to plot many types of 'linear' 2-dimensional graphs from data which is stored in vectors or matrices. The user has control over the data points to be used for a plot, the line styles and colours and the markers for plotted points. All of these variations are detailed under plot in the User's Guide. In this section we simply illustrate some of them with examples.

14.1 Markers

The markers used for points in plot may be any of

point	·
circle	o
cross	×
plus	+
star	*

● **If any of these markers is used the corresponding plot consists only of the discrete points.**

14.2 Line styles

The line style can be a solid line as before or one of several broken lines. These are particularly useful in distinguishing curves for monochrome printing. These styles are denoted by

solid line	–
dotted line	:
dashed line	– –
dash-dot	–.

14.3 Colours

The colours available for MATLAB plots are

yellow	y
green	g
magenta	m
blue	b
cyan	c
white	w
red	r
black	k

The colour choice may be combined with any of the line styles or markers.

The use of these is illustrated with the following commands. First the data vectors

```
» x=(-5:0.1:5)';
» y=twofuns(x);
```

are used for our illustrations. The file twofuns.m is

```
function y=twofuns(x)
y(:,1)=1./(1+x.^2);
y(:,2)=x./(1+x.^2);
```

The matrix y has two columns the first of which has the values of $\dfrac{1}{1+x^2}$ and the second has the values of $\dfrac{x}{1+x^2}$.

Perhaps the simplest method of obtaining multiple plots on the same axes is to use the command hold on. The effect of this is that the next plot will be added to the current plot window. To cancel this in order to produce a new figure, hold o can be used. The command hold acts as a simple toggle switch between these two states.

To obtain a plot of just the first of these functions, we can use

```
» plot(x,y(:,1))
```

which produces a 'blue solid line' graph like that below

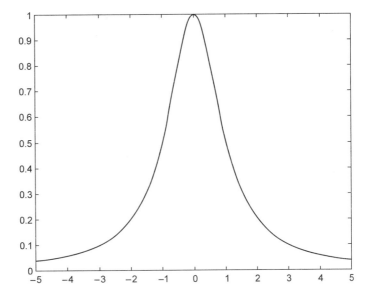

Blue, solid line is the default first graph generated by plot. To change this we use syntax such as 'm+' which would be added to the command to force magenta plus signs to be plotted. The command

```
» plot(x,y(:,1),'k:')
```

yields the same curve plotted as a dotted black line.

It is worth noting that these effects are reliable screen effects, and are usually reliable for printing directly from MATLAB. Sometimes when graphics such as these are copied into other documents, the printer drivers will convert all line styles to solid lines!

To get the plots of both functions on the same axes – but with different linestyles – use a command such as

```
≫ plot(x,y(:,1),'k',x,y(:,2),'g- -');
```

which results in the first curve being plotted as a solid black line, and the second as a dashed green line.

Note that the vector x **is needed for both sets of data.**

15 MATLAB program control

We consider the various ways of controlling the *flow* of a piece of MATLAB code. These include:

- **for** loops – which enable us to have an operation repeated a specified number of times.
  This may be required in summing terms of a series, or specifying the elements of a nonuniformly spaced vector such as the first terms of a sequence defined recursively.

- **while** loops – which allow the number of times the loop operation is performed to be determined by the results.
  This is often used in iterative processes such as obtaining approximations to the solution of an equation.

- **if . . . else . . .** – which is used when different actions are to be taken depending on the value of some variable.
  A simple example would be in obtaining the maximum of 2 quantities.

Often the else clause may not be needed when a command is to be performed only if some condition is satisfied while *nothing* is done if the condition is not satisfied. The *condition* itself may be a compound condition based on more than one 'true-false' determination. We shall see examples of all these.

15.1 for loops

The simplest for loops in MATLAB are of the form

```
for var=start : finish
    first_command
    . . .
    last_command
end
```

where start and finish are integers usually with start < finish. (If finish < start, the loop is said to be *empty* and none of the commands will be performed.)

Each of the commands in the loop will be performed once for each value of var beginning with the start value and increasing by 1 each time until the commands are executed for the last time with var=finish.

As a simple example the following commands generate the sum

$$2 + 2/3 + 2/9 + \cdots + 2/3^{10}$$

The terms are generated recursively – each is just one-third of its predecessor:

```
» term=2;
» S=2;
» for k=1:10
    term=term/3;
    S=S+term;
end
» S
S =
    2.99998306491219
```

Note that both term and sum must be initialized before the loop.

for loops can be nested. For example if we wished to repeat the summation for common ratios other than 3, we might use the following loops:

```
» for n=2:5
    term=2;
    S(n)=2;
    for k=1:10
        term=term/n;
        S(n)=S(n)+term;
    end
end
[(2:5)',S(2:5)']
ans =
    2 3.99805
    3 2.99998
    4 2.66667
    5 2.50000
```

for loops can use a step other than 1. The general format is

```
for counter=start:step:stop
    . . .
end
```

15.2   while loops

For many situations a loop is required *but* we don't know in advance how many times its commands will need to be performed. One situation where this arises regularly is in obtaining repeated approximations to a quantity where we seek to achieve some specified accuracy in the final result. For example, if we set $x = 1$ and then repeat the instruction

```
» x=cos(x)
```

we will see that the values begin to settle down. Suppose we wish to compute these until they agree to within 4 decimal places.

To achieve this, we can use a while loop which is controlled not by a simple counter but by a logical condition. The loop

```
» x1=1;x0=0;
» while abs(x1-x0)>1e-4
    x0=x1;
    x1=cos(x0);
end
```

yields the final result

```
x1 = 0.7391
```

Note that we need 2 values in order to make the comparison. Therefore 2 initial values must be specified, and we must update them each time through the loop.

Again, while loops can be nested, while loops may contain for loops, and vice versa.

The control of while loops can be more complicated than here. It can contain compound logical conditions.

16   Logical and relational operators

There are three basic logical operators which are important in MATLAB programming:

```
&    AND
|    OR
~    NOT
```

The meanings of these should be fairly obvious. Just remember that the mathematical meaning of OR is *always* inclusive. The effects of these operators is simply summarized in the following *truth table* where T and F denote 'true' and 'false' respectively. In MATLAB these values are actually represented numerically using 0 for false and 1 for true. (Strictly, any *nonzero* equates to 'true' so that some arithmetic can be done with logical values – *if you are very careful*!)

A	B	A&B	A|B	~A
T	T	T	T	F
T	F	F	T	F
F	T	F	T	T
F	F	F	F	T

There are other logical *functions* in MATLAB. The details can be found in the User's Guide. These include the *exclusive* or, xor, and all and any which have fairly obvious meaning for a logical test on vector data. (We shall not use these extensively in this book.)

Generally, you should use ( ) round the various components of compound logical conditions to be sure you have the precedence you intend. For example, (A&B)|C and A&(B|C) have very different meanings – just compare the two truth tables.

The relational operators which are often used in testing are mostly straightforward:

<	Less than
>	Greater than
<=	Less than or equal to
>=	Greater than or equal to
==	Equal to
~=	Not equal to

Note especially the *double* == for logical testing of equality. The single = is used in MATLAB only for assignments.

17 if . . . else . . .

Finally, we introduce the basic structure of MATLAB's logical branching commands. Frequently in programs we wish the computer to take different actions depending on the value of some variables. Strictly these are logical variables, or, more commonly, logical *expressions* similar to those we saw in defining while loops.

There are three basic constructions all of which begin with if and finish with end. The simplest has the form

if *condition*
    commands
end

in which the statements in the *commands* block are executed *only* if the *condition* is satisfied (true). If the condition is not satisfied (false) then the *commands* block is skipped.

The second situation is

if *condition*
    true_commands
else
    false_commands
end

in which the first set of instructions, the *true_commands* block, are executed if *condition* is true while the second set, the *false_commands* block, are executed if *condition* is false.

As a simple example the following code finds the maximum of two numbers:

```
» if a>=b
     maxab=a;
  else
     maxab=b;
end
```

The third, and most general, form of branch is

if *first_condition*
    first_true_commands
elseif *second_condition*
    second_true_commands
else
    second_false_commands
end

In this case, the *first_true_commands* block is executed if the *first_condition* is true. Otherwise (in which case *first_condition* is false) a *second_condition* is tested: if it is true the *second_true_commands* block is executed; if this second condition is *also* false then the *second_false_commands* block is executed.

In fact, if ... elseif ... elseif ... ··· ... else ... end blocks can be *nested* arbitrarily deep. Only the statements associated with the *first* true condition will be executed.

The code below could be used to determine the nature of the roots of an arbitrary quadratic equation $ax^2 + bx + c = 0$.

```
» if a==0
     if b==0
       if c==0
          fprintf('Every x is a solution \n')
       else                % else of c==0
          fprintf('No solutions \n')
       end
     else                % else of b==0
       fprintf('One solution \n');
     end
     elseif b^2-4*a*c>0        % else of a==0,
       fprintf('Two real solutions \n');
     elseif b^2-4*a*c==0
       fprintf('One repeated real solution \n');
     else
       fprintf('Two complex solutions \');
end
```

B Answers to Selected Exercises

Chapter 1

Section 1.2

1 (a) Chopping: $e = 2.718 \times 10^0$, Rounding: $e = 2.718 \times 10^0$
(b) $e = 2.718281 \times 10^0$; $e = 2.718282 \times 10^0$
(c) $e = 1.010110111 \times 2^1$; $e = 1.010111000 \times 2^1$

2 $1/3 = (0.010101\ldots)_2 = 1.01010\ldots \times 2^{-2}$
$1/5 = (0.00110011\ldots)_2 = 1.1001100\ldots \times 2^{-3}$
$1/6 = (0.0010101\ldots)_2 = 1.01010\ldots \times 2^{-3}$

Section 1.3

3 Tail for $N = 4$ is approximately 10^{-7}, for $N = 5$ it is smaller than 10^{-9} so
5 terms suffice

4 Truncation error is bounded by $|x|^7[8/(8 - |x|)]/7!$. For small x, $8 - x \approx 8$ so
we require $|x|^7 < 7! \times 10^{-10}$, or $|x| < 0.12599$

Section 1.4

2 Absolute error: $0.00003333\ldots = (1/3) \times 10^{-4}$.
Relative error: $(1/3) \times 10^{-4}/(1/3) = 10^{-4}$.

4 $\|\exp x - (1 + x)\|_\infty = e - 2 = 0.718282$. $\|\exp x - (1 + x)\|_1 = 0.218282$.
$\|\exp x - (1 + x)\|_2 = 1.67618$

Section 1.5

1 Chopping: $\hat{x} = 1.357 = \hat{y}$
$x + y = 1.3576 + 1.3574 = 2.715$;
$\hat{x} + \hat{y} = 2.714$: Relative error $= 3.6832 \times 10^{-4}$
$x - y = 1.3576 - 1.3574 = 0.0002$;
$\hat{x} - \hat{y} = 0$: Relative error $= 1$
$x \times y = 1.3576 \times 1.3574 = 1.8428$;
$\widehat{xy} = 1.841$: Relative error $= 9.7677 \times 10^{-4}$
$x/y = 1.3576/1.3574 = 1.0001$; $\widehat{x/y} = 1.00$: Relative error $= 1 \times 10^{-4}$

Chapter 2

Section 2.2

1 Polynomials are continuous, $p(0) = 4$, $p(1) = -2$, so there is a root in $[0, 1]$
Three iterations of bisection show $s \in [5/8, 3/4]$
Each iteration reduces the interval by a factor of one half. After 20 iterations the length will be reduced to $2^{-20} < 10^{-6}$

3 $f^{(3)}$ is positive everywhere, and so there are at most 3 solutions. Also $f(-1) < 0, f(0) > 0, f(1) < 0, f(9) > 0$ so there are 3 sign changes and therefore 3 solutions. Intervals: $[-0.1250, -0.0625]$, $[0.0625, 0.1250]$, $[8.9375, 9.0000]$

Section 2.3

1 (*i*)   0.85034, 1.39070, 1.93602, 1.99964, 2.00000
   (*ii*) 0.64475, 0.68139, 0.65691, 0.67320, 0.66232

2 (*ii*) converges to required solution. $s = 0.666667$

4 (*i*) $g(x) > 0$ for all x and so can only converge to a positive solution; $g'(x) = e^{x/2}/20 > 1$ if $x > 6$ so cannot converge to large solution. $g'(x) < 1$ for $x \in [0, 1]$ and so it converges to small positive solution, 0.105412
   (*ii*) Converges to large positive solution, 8.999510
   (*iii*) Converges to negative solution, -0.095345

Section 2.4

1 0.666464339908953, 0.666666660534075, 0.666666666666667, 0.666666666666667

4 For x between the 2 small solutions, $f(x) > 0$ and so $x_1 > x_0$ if and only if $f'(x_0) > 0$. Also $f'(x_0) < 0$ if x lies between the zeros of f', the smaller of which is close to 0.005. Convergence to the negative solution for $x_0 \leq 0.0049$. For x_0 very close to 0.005, the initial correction is large and eventual convergence to the large positive solution results. For $x_0 \in [0.0058, 7.28]$, we get convergence to the solution near 0.1.

6 $x_{n+1} - 1/c = -cx_n^2 + 2x_n - 1/c = -c(x_n - 1/c)^2 < 0$ so $x_{n+1} < 1/c$; also, if $x_n < 1/c$, then $x_{n+1} = x_n(2 - cx_n) > x_n(2 - 1)$

Section 2.5

5 0.16, 0.136, 0.14368, 0.14289664.

Section 2.6

1   $x^2y^3 = 1$ yields $x^2 = 1/y^3$ and substituting this in $4x^2 + y^2 = 4$ we get
    $4/y^3 + y^2 = 4$, or, multiplying by y^3, $y^5 - 4y^3 + 4 = 0$
3   First 2 iterations starting at $(0.8, 0.8)$ are $(0.76139, 0.73167)$ and
    $(0.76217, 0.71971)$
4   Intersections at $(0.76224199, 0.71943567)$ and $(-0.71472606, -1.13506001)$

Chapter 3

Section 3.2

1   Approximation: 0.223136 (True value: 0.223144 to 6 decimals)
2   8 terms, 0.2231432 (True value: 0.2231436 to 7 decimals)
4   7 terms, 1.10517091805556 (True value: 1.10517091807565)

Section 3.3

1   $0.12345 \approx 1 - \frac{1}{2} - \frac{1}{4} - \frac{1}{8} - \frac{1}{16} + \frac{1}{32} = 0.09375$.
    Error is $|0.12345 - 0.09375| = 0.0297 < \frac{1}{32}$
2   $\sigma_{k+1} + \sigma_{k+2} + \cdots + \sigma_n = \sigma_k\left(\frac{1}{2} + \frac{1}{4} + \cdots + \frac{1}{2^{n-k}}\right) = \sigma_k\left(1 - \frac{1}{2^{n-k}}\right) = \sigma_k - \sigma_n$
4   Since $1.23 \in (1, 2)$ and the error is bounded by $x_0 2^{-n}$, we require $n = 8$. With
    $x_0 = 1.23$ and $z_0 = 1.12$, we get

$$
\begin{array}{lll}
\delta_0 = +1 & y_1 = 1.23 & z_1 = 0.12 \\
\delta_1 = +1 & y_2 = 1.845 & z_1 = -0.38 \\
\delta_2 = -1 & y_3 = 1.5375 & z_1 = -0.13 \\
\delta_3 = -1 & y_4 = 1.38375 & z_1 = -0.005 \\
\vdots & \vdots & \vdots \\
& y_9 = 1.378945 &
\end{array}
$$

5   For the proof observe that $2\arctan 2^{-(k+1)} > \arctan 2^{-k}$ (take tan of both sides,
    use the formula for $\tan 2\theta$ and the fact that $\tan\theta$ is increasing for $0 < \theta \le 1$.)
    $\cos 0.5 \approx 0.88981$, $\sin 0.5 \approx 0.45598$
8   $\ln 1.5 \approx 0.41606$

Chapter 4

Section 4.1

1   1.98828125

Section 4.2

1   $(x-2)(x-4) - (x-1)(x-4) + \frac{1}{6}(x-1)(x-2)$, $59/24 = 2.4583$

3   Subtracting row k from row j yields a row of the determinant in which each element has the factor $(x_j - x_k)$. Hence the determinant is zero if any 2 nodes are equal. (To prove the converse implication, suppose the determinant is zero. Then there is a linear combination of the columns which is identically zero. This represents a polynomial of degree N which vanishes at all the $N+1$ nodes. Such a polynomial has at most N distinct roots, and so 2 of the nodes must be equal.)

4   Both values are 0.9902 to 4 decimals

7   Since $\left| f^{(4)}(x) \right| \le 1$ for all x, the error is bounded by $|(x-x_0)(x-x_1)(x-x_2)(x-x_3)|/24$. Here x lies between x_1 and x_2, and writing $x = x_0 + sh$ with $h = 0.1$, the error is bounded by $10^{-4} s(s-1)(s-2)(s-3)/24$. The maximum value of this function for $s \in [1, 2]$ occurs at $s = 1.5$ for which this error bound is $10^{-4}(1.5)(0.5)(-0.5)(-1.5)/24 \approx 2.34 \times 10^{-6}$.

Section 4.3

1   $f[x_0, x_1] = 1, f[x_0, x_2] = 2/3, f[x_0, x_1, x_2] = -1/6$

2   $p(x) = 1 + x - x(x-1)/63$

3   Using the points in the order $0.2, 0.3, 0.4, 0.0, 0.6$, the successive approximations are

$$6.232, 6.232 + (0.04)(1.030) = 6.2732$$
$$6.2732 + (0.04)(-0.06)(-0.15) = 6.2736$$
$$6.2736 + (0.04)(-0.06)(-0.16)(-0.16667) = 6.2735$$
$$6.2735 + (0.04)(-0.06)(-0.16)(0.24)(0.41667) = 6.2735$$

4   0.9902, 0.9394, 0.7776

7   (b)   Using forward difference formula:

$$p(x) = 1 + x(-1) + \frac{x(x-1)}{2!}(2) + \frac{x(x-1)(x-2)}{3!}(-2)$$

8   Induction step: suppose true for $n = m$ and consider $n = m + 1$. By definition

$$
\begin{aligned}
&f[x_k, x_{k+1}, \ldots, x_{k+m}, x_{k+m+1}] \\
&= \frac{f[x_{k+1}, \ldots, x_{k+m}, x_{k+m+1}] - f[x_k, x_{k+1}, \ldots, x_{k+m}]}{(m+1)h} \\
&= \frac{\Delta^m f(x_{k+1})/h^m m! - \Delta^m f(x_k)/h^m m!}{(m+1)h} \\
&= \frac{\Delta^m f(x_{k+1}) - \Delta^m f(x_k)}{h^{m+1}(m+1)!} = \frac{\Delta^{m+1} f(x_k)}{h^{m+1}(m+1)!}
\end{aligned}
$$

Section 4.4

1   (*a*)   Not a spline, (*b*)   Spline of degree 1, (*c*)   Spline of degree 2,
     (*d*)   Not a spline, (*e*)   Spline of degree 3
2   Approximations: $f(2) \approx 3.5, f(3.5) \approx 2.5, f(4.5) \approx 1.75$
4   $c_0 = 0, c_1 = -0.2093, c_2 = -0.0255, c_3 = -0.0421$. Then
     $b_0 = 0.7629, b_1 = 0.5535, b_2 = 0.3187, b_3 = 0.2511$ and
     $d_0 = -0.0698, d_1 = 0.0613, d_2 = -0.0055, d_3 = 0.0140$
6   Since $c_0 = 0$, $a_0 + b_0(x - x_0)$ has same value, and first and second derivatives
     as s_0 at $x = x_0$; for s_n apply equation (4.30) with $d_n = c_n = 0$.

Chapter 5

Section 5.2

2   $\frac{4}{3}[2f(-1) - f(0) + 2f(1)]$; note that quadrature rules can have negative
     coefficients!
3   Quadrature rule is $\frac{2}{45}[7f(-2) + 32f(-1) + 12f(0) + 32f(1) + 7f(2)]$
     Estimate of area of semicircle is 5.9934
4   $M = 0.8$, $T = 0.75$, $S = 0.783333$, $\pi/4 = 0.785398$
5   $M = 0.66667$, $T = 0.75$, $S = 0.69444$, $\ln 2 = 0.693147$
6   Nodes: $\frac{1}{2} \pm \frac{\sqrt{3}}{2\sqrt{5}}, \frac{1}{2}$ with weights $\frac{5}{18}, \frac{4}{9}$. 0.78527, 0.69312
8   Nodes: $\pm 0.861136, \pm 0.339981$ with corresponding weights 0.347855,
     0.652145. Integral ≈ 1.56863

Section 5.3

1   1, 1.0667, 1.0898, 1.0963, 1.0980
6   $M = 0.97743$, $T = 0.97689$, $S = 0.97724887$
7   $N = 20$: $M = 0.97729$, $T = 0.97716$, $S = 0.97724981$
     $N = 40$: $M = 0.97726$, $T = 0.97723$, $S = 0.97724986$
     $N = 80$: $M = 0.977253$, $T = 0.977244$, $S = 0.97724987$
     $N = 160$: $M = 0.977251$, $T = 0.977248$, $S = 0.97724987$

Section 5.4

1   For $\int_0^1 1/(1+x^2)dx$, we get $T_1 = (1 + 1/2)/2 = 3/4$,
     $T_2 = (1 + 2 \times 4/5 + 1/2)/4 = 31/40$, $S_2 = (1 + 4 \times 4/5 + 1/2)/6 = 47/60$,
     and $(4T_2 - T_1)/3 = (31/10 - 3/4)/3 = 47/60$, also
3   $T_1 = \dfrac{b-a}{2}(f(a) + f(b))$, $M_1 = (b-a)f(m)$ where $m = (a+b)/2$.

     $T_1 + M_1 = \dfrac{(b-a)}{2}(f(a) + 2f(m) + f(b)) = 2T_2$

4 1.2000   0   0   0
   1.1000   1.0667   0   0
   1.1038   1.1051   1.1077   0
   1.1063   1.1071   1.1073   1.1073

5 1.57079633

7 $N = 5 : 1.5707963267948$, $N = 10 : 1.5707963267948$,
$N = 20 : 1.5707963267948$

Section 5.5

1 0.21938392

2 0.278806

3 0.27880551, 0.21938392, 0.17814771, 0.14849551, 0.12648783, 0.10969198,
0.09655666, 0.086062506, 0.07752082, 0.07045426, 0.06452417, 0.05948508

Section 5.6

1 (*a*) 0.488088, 0.498756, 0.499875, 0.4999875, 0.49999875, 0.499999875
   (*b*) 4.1, 4.01, 4.001, 4.0001, 4.00001, 4.000001
   (*c*) 0.953102, 0.995033, 0.999500, 0.999950, 0.999995, 0.9999995

2 (*a*) 0.513167, 0.501256, 0.500125, 0.5000125, 0.50000125, 0.500000125
   (*b*) 3.9, 3.99, 3.999, 3.9999, 3.99999, 3.999999
   (*c*) 1.053605, 1.005034, 1.000500, 1.000050, 1.000005, 1.0000005

3 **Hint**: Start from the central difference formula (4.18)

4 (*a*) 0.500628, 0.500006, 0.50000006, 0.5000000006, 0.500000000003,
      0.50000000001
   (*b*) 4, 3.99999999999994, 3.99999999999956, 4.000000000004,
      4.0000000000262, 4.00000000011502
   (*c*) 1.003353, 1.000033, 1.00000033, 1.0000000033, 1.000000000034,
      0.99999999997

6 (*a*) 0.4999887, 0.4999999989, 0.4999999999998, 0.4999999999992,
      0.499999999996, 0.499999999996
   (*b*) 4, 3.99999999999992, 3.99999999999963, 4.00000000000696,
      4.0000000000262, 4.00000000011502
   (*c*) 0.999917, 0.999999992, 0.9999999999991, 0.99999999999989,
      1.00000000000192, 0.999999999963993

8 Using second-order formula
   (*a*) −0.250785, −0.2500078, −0.250000078, −0.2500000096,
      −0.250000021, −0.249911
   (*b*) 2, 1.9999999999909, 2.00000000027956, 2.000000077, 2.00000017,
      2.001066
   (*c*) −1.005034, −1.00005000, −1.000000500, −1.0000000050, −0.9999989,
      −1.000111

Using fourth-order formula
(*a*) −0.249983, −0.249999998, −0.24999999991, −0.250000015,
       −0.25000002, −0.249893
(*b*) 1.99999999999997, 1.99999999999016, 2.0000000004, 2.000000088,
       1.9999965, 2.001325
(*c*) −0.999861, −0.999999987, −1.0000000001, −0.9999999999997,
       −0.9999984, −1.000157

Section 5.7

1 $[1.1, 1.7]$
2 $[1.3, 1.7]$
3 First iteration: $1.3, 1.4319, 1.7$
4 1.385224
5 First iteration: $1.3, 1.4020$

Chapter 6

Section 6.1

1 Separate variables: $y^2 - x^2 = c$
2 For $h = 1, 1/2, 1/4$ estimates of $y(1)$ are 3, 3.0833, 3.1234
3 Estimates of $y(4)$ are 4.8847, 4.9434, 4.9776, 4.9888, 4.9944

6

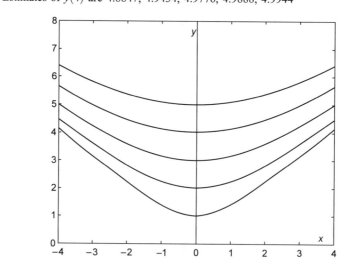

Section 6.2

1 For $h = 1, 1/2, 1/4$ estimates of $y(1)$ are 3.1667, 3.1633, 3.1625 (True solution is $\sqrt{10} = 3.162\,277\,7$.)

2 Modified Euler: 3.1667, 3.1628, 3.1623
Heun: 3.1667, 3.1631, 3.1625

3 Estimates of $y(4)$:

N	CorrEuler	ModEuler	Heun
10	5.0027	5.0003	5.0019
20	5.0007	5.0000	5.0005
50	5.0001	5.0000	5.0001
100	5.0000	5.0000	5.0000

8 For $N = 10, 20, 50, 100$, estimates of $y(4)$ are 5.00000203217189, 5.00000012217172, 5.00000000305775, 5.00000000018970
Errors are $2.0 \times 10^{-6}, 1.2 \times 10^{-7}, 3.0 \times 10^{-9}, 1.9 \times 10^{-10}$

Section 6.3

1

x	1/4	1/2	3/4	1
y	3.0104	3.0416	3.0928	3.1632

2 Values at $x = 1$: AB2: 5.00318, Modified Euler: 5.00032, Corrected Euler: 5.00271

3 Values at $x = 1$ for 20, 40, 100, 200 steps: 5.00318, 5.00080, 5.00013, 5.00003
Errors: 0.00318, 0.00080, 0.00012, 0.000032; supports the claim that error is $O(h^2)$

4 At $x = 1$, 4.99711, error $= 3 \times 10^{-3}$

5 At $x = 1$, 4.99977, error $= 2.3 \times 10^{-4}$

6

N	Error(ABM22)	Error(ABM23)
10	0.002889	2.3×10^{-4}
20	0.000687	3.5×10^{-5}
50	0.000105	2.4×10^{-6}
100	0.000026	3.1×10^{-7}

9 AB4 formula: $y_{n+1} = y_n + \dfrac{h}{24}(55f_n - 59f_{n-1} + 37f_{n-2} - 9f_{n-3})$

12

N	ABM34	ABM44	RK4 with $N/2$
20	4.2×10^{-6}	3.6×10^{-6}	2.0×10^{-6}
40	2.7×10^{-7}	2.2×10^{-7}	1.2×10^{-7}
80	1.7×10^{-8}	1.4×10^{-8}	7.5×10^{-9}

Section 6.4

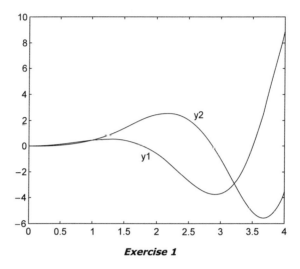

Exercise 1

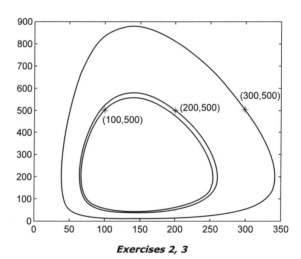

Exercises 2, 3

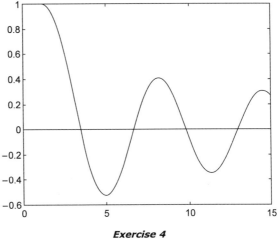

Exercise 4

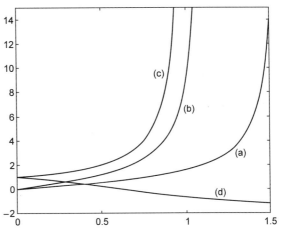

Exercise 5

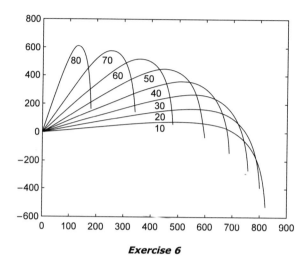

Exercise 6

Section 6.5

1   Initial condition needed is $y'(1) = 1.5020$
2   Initial condition needed is $y'(1) = 4.6958$
3   Required initial conditions: (*a*) $y'(0) = 0.7402$, (*b*) $y'(0) = 1.9650$,
   (*c*) $y(0) = -0.6736$
4   Approximate launch angle 18.16^0.

Section 6.6

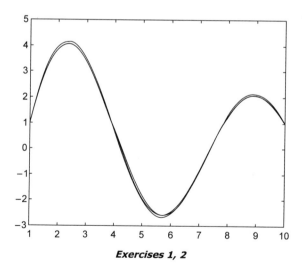

Exercises 1, 2

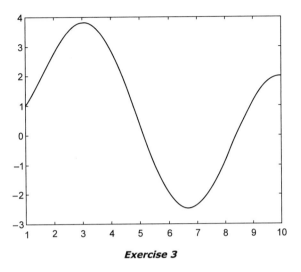

Exercise 3

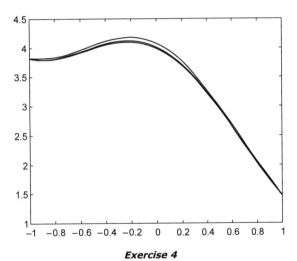

Exercise 4

Section 7.2

1   $[w, x, y, z] = [1, -1, 1, 2]$

4   In short format results are all ones until $n = 10$ for which we get $[1, 1, 1, 1, 0.9999, 1.0003, 0.9996, 1.0004, 0.9998, 1]$. Long format output shows errors gradually getting worse as dimension increases

5   Same as for Exercise 4

7

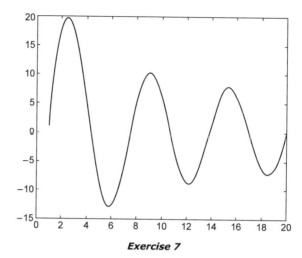

Exercise 7

Section 7.3

1

$$L = \begin{bmatrix} 1 & & \\ 0.8571 & 1 & \\ 0.1429 & -0.4614 & 1 \end{bmatrix}, \qquad U = \begin{bmatrix} 7 & 8 & 8 \\ & -1.8568 & -2.8568 \\ & & 0.5387 \end{bmatrix}$$

Initial solution $\tilde{\mathbf{x}} = \begin{bmatrix} 0.9998 \\ -0.9992 \\ 0.9994 \end{bmatrix}$ Residues $\begin{bmatrix} -0.0002 \\ -0.0004 \\ 0.0004 \end{bmatrix}$.

Improved solution $\mathbf{x} = \begin{bmatrix} 1 \\ -1 \\ 1 \end{bmatrix}$

2 Solution is $[0.1, 0.3, 0.5, 0.7, 0.9]'$

3, 4 For $n = 10$, the initial solutions and the results of iterative refinement are

1.00000000033336	1.00000000062242
0.999999969559204	0.999999949045136
1.00000067719383	1.00000103731161
0.999993626315953	0.999990926237516
1.00003126880402	1.00004186451376
0.999912025222801	0.999888206781766
1.00014716762512	1.00017878783228
0.999855421456933	0.999831102392846
1.0000769803009	1.00008688636632
0.999982862835067	0.999981238666823

5 Solution vector is approximately $[-3.4607, 10.1363, -12.1628, 8.0199, -3.2586, 0.8603, -0.1505, 0.0173, -0.0013, 0.0001]$

Section 7.4

1   Jacobi iterations: $[1.0250, 0.6000, 1.0500, 0.6000, 1.0250]$;
    $[0.8750, 0.0812, 0.7500, 0.0812, 0.8750]$;
    $[1.0047, 0.1937, 1.0094, 0.1937, 1.0047]$

2   Gauss–Seidel iterations: $[1.0250, 0.3438, 0.9641, 0.3590, 0.9353]$;
    $[0.9391, 0.1242, 0.9292, 0.1339, 0.9915]$;
    $[0.9939, 0.1192, 0.9867, 0.1054, 0.9986]$ which is much closer to the true
    solution

Section 7.5

1   $15/2\pi^2(1 - 3x^2)$

3   Integrals are $0, 0, -4/\pi^2, 0, 8(6 - \pi^2)/\pi^4$. Therefore the coefficients of
    $P_0, \ldots, P_4$ are $0, 0, -1.51982, 0, 0.58245$

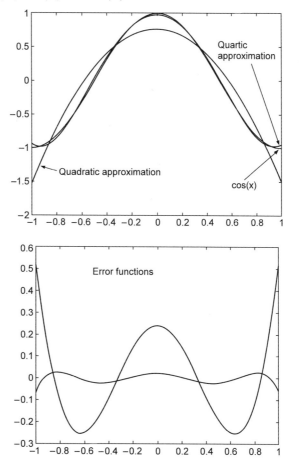

4 $-2.1039x^2 + 0.7238$

5 $2.4543x^4 - 4.4005x^2 + 0.9737$

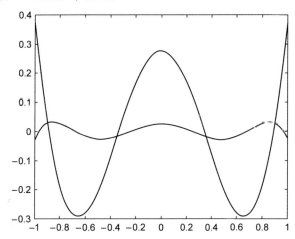

Exercise 4,5   Error curves

7 $1, x - 1/2, x^2 - x + 1/6, x^3 - 3x^2/2 + 3x/5 - 1/20,$
$x^4 - 2x^3 + 9x^2/7 - 2x/7 + 1/70$

8 Coefficients: 0.4388, 0.7918, −0.2886, −0.0603, 0.1415

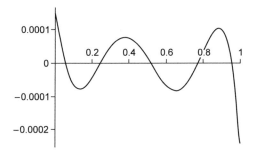

Exercise 8   Error curve

Section 7.6

1   25.69815

2   0.26821

3 Inverse iteration code:

```
function lammin=invit(A,x0,tol)
% Inverse iteration for finding |smallest| eigenvalue
[L,U]=lu(A);
Lam0=-inf; Lam1=inf; its=0;
while abs(Lam0-Lam1)>tol
   Lam0=Lam1; its=its+1;
   w=L\x0; v=U\w;
   Lam1=dot(v,x0)/dot(x0,x0);
   x0=v/norm(v);
end
lammin=1/Lam1;
```

4.90635

4 -1.04490

5 0.50706

6 -0.33487

References and Further Reading

Barlow, J.L. and Bareiss, E.H. (1985) 'On roundoff error distributions in floating-point and logarithmic arithmetic', *Computing,* **34**, 325–47.

Buchanan, J.L. and Turner, P.R. (1992) *Numerical Methods and Analysis*, New York: McGraw-Hill.

Burden, R.L. and Faires, J.D. (1993) *Numerical Analysts*, 5th edn, Boston: PWS-Kent.

Cheney, E.W. and Kincaid, D. (1994) *Numerical Mathematics and Computing*, 3rd edn, Pacific Grove, CA: Brooks/Cole.

Clenshaw, C.W. and Curtis, A.R. (1960) 'A method for numerical integration on an automatic computer', *Numerische Mathematik,* **2**, 197–205.

Clenshaw, C.W. and Olver, F.W.J. (1984) 'Beyond floating-point', *Journal of the ACM,* **31**, 319–28.

Davis, P.J. and Rabinowitz, P. (1984) *Methods of Numerical Integration*, 2nd edn, New York: Academic Press.

Feldstein, A. and Goodman, R. (1982) 'Loss of significance in floating-point subtraction and addition', *IEEE Transactions on Computers,* **31**, 328–35.

Feldstein, A. and Turner, P.R. (1986) 'Overflow, underflow and severe loss of precision in floating-point addition and subtraction', *IMA Journal of Numerical Analysis,* **6**, 241–51.

Hamming, R.W. (1970) 'On the distribution of numbers', *Bell Systems Technical Journal,* **49**, 1609–25.

Kincaid, D. and Cheney, E.W. (1991) *Numerical Analysis*, Pacific Grove, CA: Brooks/Cole.

Knuth, D.E. (1969) *The Art of Computer Programming, Vol.2, Seminumerical algorithms*, Reading, MA: Addison-Wesley.

MathWorks, The (1995) *The Student Edition of MATLAB, Version 4 User's Guide*, Englewood Cliffs, NJ: Prentice-Hall.

MathWorks, The (1997) *The Student Edition of MATLAB, Version 5 User's Guide*, Englewood Cliffs, NJ: Prentice-Hall.

Olver, F.W.J. (1978) 'A new approach to error arithmetic', *SIAM Journal of Numerical Analysis,* **15**, 369–93.

Pratap, R. (1999) *Getting Started with MATLAB 5*, Oxford: Oxford University Press.

Schelin, C.W. (1983) 'Calculator function approximation', *American Mathematical Monthly,* **90**, 317–25.

Turner, P.R. (1982) 'The distribution of leading significant digits', *IMA Journal of Numerical Analysis,* **2**, 407–12.

Turner, P.R. (1984) 'Further revelations on l.s.d.', *IMA Journal of Numerical Analysis,* **4**, 225–31.

Volder, J. (1959) 'The CORDIC computing technique', *IRE Transations on Computers,* **8**, 330–4.

Walther, J. (1971) 'A unified algorithm for elementary functions', *AFIPS Conference Proceedings,* **38**, 379–85.

Wilkinson, J.H. (1963) 'Rounding errors in algebraic processes', *Notes on Applied Science*, HMSO, London.